STEAM GENERATION FROM BIOMASS

STEAM GENERATION FROM BIOMASS

Construction and Design of Large Boilers

ESA KARI VAKKILAINEN

Lappeenranta University of Technology, Lappeenranta, Finland

AMSTERDAM • BOSTON • HEIDELBERG • LONDON • NEW YORK • OXFORD
PARIS • SAN DIEGO • SAN FRANCISCO • SINGAPORE • SYDNEY • TOKYO
Butterworth-Heinemann is an imprint of Elsevier

Butterworth-Heinemann is an imprint of Elsevier
The Boulevard, Langford Lane, Kidlington, Oxford OX5 1GB, United Kingdom
50 Hampshire Street, 5th Floor, Cambridge, MA 02139, United States

British Library Cataloguing-in-Publication Data
A catalogue record for this book is available from the British Library

Library of Congress Cataloging-in-Publication Data
A catalog record for this book is available from the Library of Congress

ISBN: 978-0-12-804389-9

For Information on all Butterworth-Heinemann publications
visit our website at https://www.elsevier.com

Working together
to grow libraries in
developing countries

www.elsevier.com • www.bookaid.org

Publisher: Joe Hayton
Acquisition Editor: Raquel Zanol
Editorial Project Manager: Ana Claudia Abad Garcia
Production Project Manager: Mohanambal Natarajan
Cover Designer: Matthew Limbert

Typeset by MPS Limited, Chennai, India

CONTENTS

The companion materials for this book are available at
http://booksite.elsevier.com/9780128043899/

ABOUT THE AUTHOR

Esa Vakkilainen is Professor of Sustainable Energy Systems at Lappeenranta University of Technology, Finland.

His professional career started in the 1980s as an assistant professor of power plants, teaching and researching furnace heat transfer, optimization of combined cycle processes, district heating (a Finnish specialty), and the combustion of biofuels.

After 4 years in academia, he joined A. Ahlstrom Corporation and started working at their Varkaus boiler works. There he was responsible for developing a new generation of dimensioning programs for steam generator thermal design. It was a time of extensive development as circulating fluidized beds (CFBs) were making their way into the mainstream of steam power plants. After several changes of ownership, these boilers are now manufactured by the AMEC Foster Wheeler Corporation.

In addition to working on CFBs, Esa Vakkilainen has been involved (as research manager and technical manager) with biomass boilers such as kraft recovery boilers. His main interests have been the fouling of heat transfer surfaces, fuel delivery, air distribution, and combustion.

Prior to taking up his current position, as a leading technology expert on energy and environment he was employed by Jaakko Pöyry Oy, international consultants to the pulp and paper industry. He has thus been involved with the majority of recent purchases of biomass boilers for the pulp and paper industry worldwide.

Esa Vakkilainen has pursued active research, which has led to the supervision of around 200 MSc and several PhD. theses. He has lectured on boilers in technical conferences in all the major continents.

PREFACE

The purpose of this book is to cover steam generator engineering for biomass combustion in depth. There has not been a widely available English language book on this subject for some time. In fact there are very few English books on the subject of steam generator design for any fuels. With more stringent emission limits and the looming increase of CO_2 in our atmosphere we are more interested in getting our power produced in the most environmentally beneficial way.

Any technical device is designed for a specific purpose and to fulfill specified operating conditions. To be able to gauge how, e.g., the choice of fuel affects the steam generator, a reader should first understand the design process. One needs to study the different components that comprise a steam generator and to understand not only their function. Finally, the reader should be able to look at a steam generator and point out what design choices have been made.

This book is intended not only for students of engineering but also for both the design engineers and operators of steam boilers. It describes steam generator manufacture, but does not cover construction details in depth. Throughout the book it is assumed that the reader has a basic understanding of the concepts used in thermodynamics, heat transfer, and fluid flow.

Each chapter of the book deals with a particular aspect of steam generator design. In the author's view this is a holistic process – the whole consists of optimal designs of several subdisciplines. Examples of such subdisciplines are pressure part design, mechanical design, layout design, process design, performance optimization, cost optimization, etc. One can, however, easily become acquainted with the functioning of a steam generator without in-depth knowledge of all of these.

I am greatly indebted to the clarifications and advice of several colleagues who took the time to proofread various versions of my copy and suggest changes. Thank you Edgardo Coda Zabetta from Amec Foster Wheeler, Sebastian Kankkonen from Elomatic, Jouni Kinni from Valmet Power, and Janne Tiihonen from Andritz.

To everyone, I say: "Enjoy your journey through this book!"

POEM

*Total is the sum of parts, to master the steam boiler one must
learn the functioning of every piece of equipment.*

> The boiler is the soul of the whole,
> the part where the power is generated,
> the engine being merely an agent
> for transmitting power from the boiler
> to the work that is to be performed.
>
> *Richards (1873)*

DISCLAIMER

While every effort has been made to ensure the accuracy of material presented in this book, the author wishes to state, that no part of this book including individual figures and examples should be accepted at face value. Even though the examples deal with steam generators as they are designed today, no assurance is given of the validity of using material in this book in the actual design.

Steam generator design is always a compromise that involves such factors as manufacturing cost, available materials, specific fuel properties, and previous practice. Design choices that seemed valid have number and number of times proven to have inherent faults when applied to different size, fuel, or operating environment.

NOMENCLATURE

A	surface area, m^2
a_d	dust emission coefficient, -
B_d	dust loading, kg/m^3
b	correction coefficient, -
C	heat capacity, W/C
c_p	specific heat capacity, kJ/kgC
d	diameter, m
d	depth, m
df	dry fuel
d_o	outside tube diameter, m
d_s	inside tube diameter, m
f	correction factor for heat transfer, -
f_n	form correction, -
f_o	overall correction, -
G	conductance, W/C
H	Heating value, J/kg
h	height, m
k	heat transfer coefficient, W/m^2C
k_c	convective heat transfer coefficient, W/m^2C
k_i	inside heat transfer coefficient referred to the outside surface, W/m^2C
k_o	outside heat transfer coefficient, W/m^2C
k_r	radiative heat transfer coefficient, W/m^2C
k_{ex}	external heat transfer as a heat transfer coefficient, W/m^2C
L	length, m
M	mole weight, kg/kmol
n	number of, -
P	mechanical energy, W
p	pressure, bar abs
p_x	is the partial pressure of substance x, bar
q_m	mass flow, kg/s
Q	heat, W
R	ratio of steam side heat capacity to gas side heat capacity, -
r	radius, m
s	radiation beam length, m, tube wall thickness, m
S_l	longitudinal pitch, m
S_t	transverse pitch, m

s_l	dimensionless longitudinal pitch, -
s_t	dimensionless transverse pitch, -
T	temperature, K or °C
w	flow velocity, m/s
w	width, m
V	flow, m³/s
X	mass fraction, -
z	number of transfer units, -
α_d	absorption coefficient of dust, -
α_{dg}	absorptivity of the dusty gas, -
ε	emissivity, -
ε	ratio of gas temperature drop to total temperature difference, -
ε_{dg}	emissivity of the dusty gas, -
ε_α	background emissivity, -
ε_w	emissivity of the wall, -
	loss coefficient, -
η	viscosity, Pas
η	efficiency, -
θ	total temperature difference, K or °C
λ	heat conductivity, W/m°C
ξ	friction factor, -
ρ	density, kg/m³
Δ	difference, -
ΔT	temperature difference, K or °C
$\Delta\varepsilon_g$	overlapping correction for emissivity, -
$\Delta\alpha_g$	overlapping correction for absorptivity, -
Φ	heat flow, W

Subscripts

a	air, arrangement
abs	heat absorbed
ad	adiabatic
ah	air heater
ash	ash
av	average
b	bend
bd	blowdown
c	convection
d	dry, dust, dynamic, particle
dg	dusty gas
dry	dry
dsh	desuperheat
e	exit

eco	economaiser
eff	effective
ev	evaporative
ex	external
F	furnace
f	fouling, friction, fuel
fw	feed water
g	gas
gs	gas side
ln	logarithmic
i	in, inlet, inside
in	inside
io	inlet and outlet
l	laminar, longitudinal, losses
mix	mixture
ms	main steam
net	net
o	out, outlet, outside
p	pyrolysable
pipe	tubes
plate	platen
r	radiation, row
rh	reheat
s	steam
sh	superheter
sw	side wall
t	total, transverse, turbulent, tube
tot	total
ts	tube side
v	valve
w	wall, water
wet	wet
x	index
0	initial

1

PRINCIPLES OF STEAM GENERATION

The development of the industrial age coincided with the development of heat engines (Morris, 2010). It was the steam engine that liberated man from mainly toiling the land and afforded him the means of improving his living conditions. Currently there are four main types of heat engine in use: internal combustion reciprocating engine, gas turbine, steam power, and rocket engine (Ishigai, 1999). Steam power is dominant in electricity-generating thermal power stations. The part of the process that is used to generate vapor is called steam generation.

The first steam generators developed around coal, especially to pump water away from deep coal mines. Biomass utilization for steam generation started at the same time in regions where coal was expensive. Steam generation from sustainably sourced biomass is gaining more interest as restricting global warming is gaining importance. The Intergovernmental Panel on Climate Change (2011) is of the opinion that, compared to the fossil energy baseline, one can achieve 80–90% fossil carbon dioxide (CO_2) emission reductions when using energy generated from biomass. The International Energy Agency (2014) in their world energy outlook expects bioenergy electricity generation in their 450 scenario to increase fivefold from 2012 to 2040.

1.1 Introduction

A steam generator that uses combustion as its main heat source is often called a boiler. Water that is converted to steam

Steam Generation from Biomass. DOI: http://dx.doi.org/10.1016/B978-0-12-804389-9.00001-0

is the most common fluid used in heat engines, which convert heat to work. Steam generation was actually first introduced in apparatus designed to convert heat to the work required for pumping water from mines. For process applications, heat from a steam generator can be used directly to serve the required process purposes (e.g., district heating by steam). In steam generation the heat source, combustion, and the working fluid are separated, typically by a wall of heat-resistant material (e.g., steel tubes).

Electricity generation is one of the main uses of steam generators with a turbine. We can compare currently predominant electricity-generating options in Table 1.1. It can easily be seen that none of the currently prevalent processes totally fulfills the most desired features.

Of the listed processes the most typical ones for new installations are still coal-fired boilers, natural gas combined cycle and biomass-fired steam generators. It should be pointed out that biomass is a renewable source of electricity and produces no net fossil CO_2 (IPPC, 2011).

Biomass is a primary source of food, fodder and fibre and as a renewable energy (RE) source provided about 10.2% (50.3 EJ) of global total primary energy supply (TPES) in 2008. Traditional use of wood, straws, charcoal, dung and other manures for cooking, space heating and lighting by generally poorer populations in developing countries accounts for about 30.7 EJ,

Table 1.1 Comparison of electricity Generation Options (Rogan, 1996)

Process	Cost of Electricity	Reliability/Availability Technology	Fuel	Environmental Emissions	CO_2 (t/MWh)
Most desirable	Low	High	High	Zero	Zero
Coal (pulverized fuel combustion)	Low	High	High	Low	1.0
Coal (advanced)	Medium	?	High	Lower	0.8
Natural gas (combined cycle)	Low	High	Varies	Lower	0.6—0.4
Nuclear	Medium	High	High	Zero	Zero
Biomass	Medium	High	Varies	Low	1.5
Solar/Wind	High	?	Varies	Zero	Zero
Fuel cells	High	?	Varies	Lower	0.3
Hydro	Low	High	Varies	Zero	Zero

*and another 20 to 40% occurs in unaccounted informal sectors
including charcoal production and distribution. TPES from biomass
for electricity, heat, combined heat and power (CHP), and transport
fuels was 11.3 EJ in 2008 compared to 9.6 EJ in 2005 and the share of
modern bioenergy was 22% compared to 20.6%. (IPPC, 2011).*

All CO_2 produced by biomass combustion is absorbed back
into the forests and other natural biomass, assuming sustainable practices are used.

What will the role of steam boilers be in the future? The
world energy consumption keeps on increasing. Currently about
70% of world electricity generation is done using thermal processes. Even with a shift to renewable sources, electricity generation based on steam power plants will continue to grow. In
spite of progress in renewable energy, world coal consumption
is still expected to grow (IEA, 2015). If we can significantly
increase steam generation from biomass, we have the possibility to dramatically decrease our carbon footprint.

1.1.1   History of Steam Generation

There is evidence of the use of steam for motive power as far
back as the Ancient Greeks (the first mention is by Heron in 200
BC). The Greeks and Romans used water heaters that have
many of the features of modern boilers. Their hot water boilers
were for domestic use only. The one in Fig. 1.1 (Croft, 1922) was
internally fired, made from bronze and had a water tube grate.
You placed the fuel on the grate and let it burn using a natural
draft to transfer the heat from combustion to the water inside.

After 1600 there was renewed interest in steam. The French,
English, and Russians used steam to drive water up into water
fountains. Some of the people involved with the first applications
of steam were Frenchman Denis Papin, Englishman Samuel
Morland, and Italians Galileo Galilei and Evangelista Torricelli.

1.1.1.1 Early Boilers

Boilers were first used for industrial application in England
in the 1700s, initially for pumping water from mines. These boilers had a very low efficiency, but as fuel supply was plentiful
and the duty required could not be met by manual labor, they
replaced horse-driven pumps.

The first practical steam engine was made by Thomas Savery
in 1698. Even though it produced positive power, the operation
could not be kept going for very long. One of the first commercially successful boilers was John Newcomen's boiler in 1712,

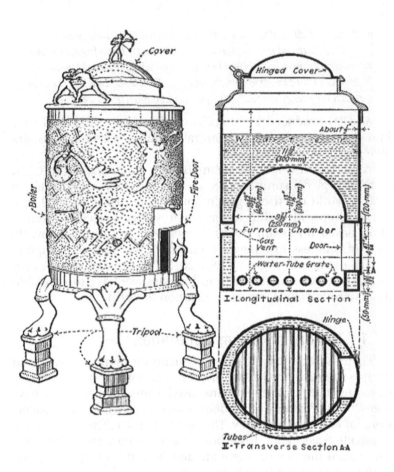

Figure 1.1 Pompeian water heater (Croft, 1922).

Figure 1.2 (Ishigai, 1999). It was the first example of a steam-driven machine capable of an extended period of operation.

This type of boiler was called the shell or spherical boiler. It was made using soldering and bent metal sheets. The Newcomen's boiler was made from copper, but iron soon replaced copper for increased pressure. Soldering was done with lead. The available steam temperature increased up to 150°C.

The first industrial biomass boiler was most probably built by Mårten Trievald in 1728 (Ångteknik, 1945), Fig. 1.3. It was one of the first outside England and was built at Dannemora Grufwor, Sweden with the intention to pump water out of the copper mines there. Because of the lack of coal, it was fired with wood (Lindqvist, 1984).

The first steam engine in the United States was built in 1754 to pump water from Schuyler Copper Mine in New Jersey (Hunter, 1985). Little is known about it except that it was built

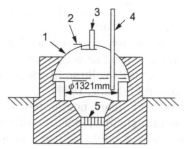

Figure 1.2 Newcomen's boiler: 1—Shell over the boiling water; 2—Steam valve; 3—Steam pipe; 4—Float for water level; 5—Grate (Ishigai, 1999).

Figure 1.3 Mårten Trievald's 1728 boiler.

from components imported from England and reportedly operated for more than half a century.

Later the spherical design was replaced by a wagon-type design for increased capacity. The wagon type also had a higher heat transfer surface per unit volume of water so efficiency was increased (Dickinson and Jenkins, 1981).

James Watt's improved steam engine led to rapid growth in steam boilers. Even though the efficiency of steam production was about 2%, these boilers could be used for extended periods of time. Boulton and Watt supplied the steam boiler and engine to Robert Foulton's 1807 famous first steamboat, *Clermont*, which started the era of commercial steamboat operation.

Steam production from boilers steadily increased. Pressure stayed low because of problems with reliable construction. One or two bars overpressure was the norm. Because of the low pressure there were few explosions resulting in fatalities.

The next development was the cylindrical steamboat boiler, which was invented by Oliver Evans and Richard Trevithick around 1800. The flue gas passed through and around two cylinders joined together, Fig. 1.4 (Forsman-Saraoja, 1928). Hence it was given the name cylindrical boiler. It had a larger heat transfer surface per unit of working fluid than the wagon boiler. Therefore the cylindrical boiler could be built more cheaply than the earlier boilers. In England these boilers were called the Lancashire and Cornish boilers.

Contaminants in the feedwater tended to settle in the bottom. For wagon boilers, which were heated by the flue gas from the bottom, this was a constant source of problems. Cylindrical boilers had the advantage that they could be built so that heat was transferred through less clogged portions of the pressure vessel. It was also easier to increase the design pressure of cylindrical boilers and in the mid-1850s the boiler pressure reached 4–8 bars. Because there was only a minor efficiency increase associated with a pressure increase when steam engines were used, the drive to increase pressure was low.

Many of the early cylindrical boilers were made of cast iron as making reliable joints was still very difficult. But development of riveting and steel making processes soon made cast iron obsolete. Cylindrical boilers were later expanded to contain several passes and eventually developed into the fire tube boiler (Lobben, 1930).

The water tube boiler was created at the end of 1800. It was first used to run the largest steam engines but it quickly became the boiler type of choice for steam turbines. The developing new industry, electricity generation, demanded boilers to drive large generators.

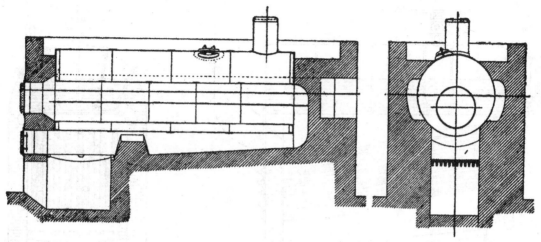

Figure 1.4 Cylindrical boiler (Forsman-Saraoja, 1928).

Water tube boilers employed a new principle. Heat in flue gas was transferred to the tubes containing water/steam by radiation and convection. The large number of tubes and the use of cross gas flow increased the available heat transfer surface and heat transfer rate. Boilers of this type could again be built for higher pressure and with a larger heat transfer surface per unit of working fluid than the fire tube boiler, Fig. 1.5 (Croft, 1922).

Wilcox, which manufactured water tube boilers, eventually merged with Babcock to form Babcock & Wilcox. The company became widely known when its boilers were employed by the Thomas Edison Company in the world's first utility electricity-generating station in Pearl Street, New York, in 1882.

1.1.1.2 Manufacturing Developments

In Wilcox's water tube boiler we can see the internal bracing put into increase the boiler's ability to withstand overpressure. Design features that aided in balancing stresses can be seen in almost all of the boilers of this era. The steel used for boiler construction started to improve dramatically as new steel manufacturing processes came into use. The Bessemer converter and Siemens-Martin open hearth steel making processes were invented around 1860. Mannesman's seamless tube manufacturing was patented in 1885. Steel prices decreased. The quality of steel improved and thicker and heavier pieces of steel could be manufactured. All this helped boiler design and manufacture.

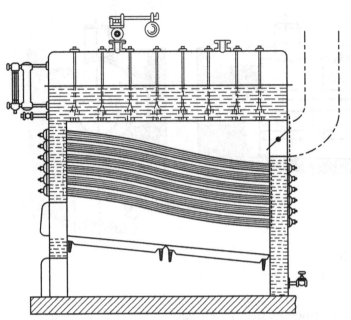
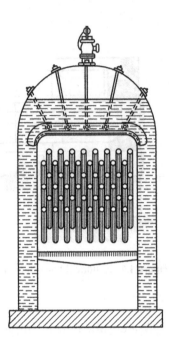

Figure 1.5 Wilcox's water tube boiler, 1856 (Croft, 1922).

Tubes could be made inexpensively and with higher quality than plate. Steel manufacture limited the practical plate thickness to 20 mm, so the use of tubes instead of sheets for boilers started to dominate. Soldering was replaced by riveting. In riveting, two sheets of metal are joined together by inserting very hot steel nuts, which are forged to form while still hot (Shields, 1961). When rivets cool down, they shrink, creating a tight seam. Riveting quickly became the main manufacturing method. Riveting remained the most used manufacturing method until the 1950s.

Another manufacturing development was the invention of forming. Here the metal is pressed with a mechanical device until the metal is permanently deformed. In boiler manufacture, forming is mostly used in the creation of rolled joints. In rolling, a tube is fastened to a drilled hole in a metal sheet by the expansion of the tube by thinning its wall. The principle employed is simple. A tube is placed in a tight-fitting hole. A special tool is inserted into the tube. By applying pressure that exceeds the steel's deformation pressure, the tube's thickness is decreased, a small section at a time. While the tube's thickness decreases, the tube's diameter increases. When the tube's outer diameter is larger than the hole's diameter, a tight join has been created. The rolling of tubes is a very successful method of low pressure joining. Rolling is successfully used even today in commercial boilers under 10 MPa.

The era of large boilers began with the invention of welding. Welding forms the basis of modern steam boiler manufacture. Welding is universally used in all modern boiler manufacture. The 1930s saw the first application of welding to boiler manufacture. With the development of X-ray technology to inspect the welds, welding became popular during the World War II.

Around 1955 the fully welded furnace (membrane wall) was developed. This meant high savings in maintenance and started rapid unit size increases.

1.1.1.3 Increasing Efficiency

The end of the 20th century saw interest in increasing the energy efficiency of boilers and steam engines. Public tests and a gradual understanding of thermodynamics changed the steam engine design criteria. Increasing the working pressure increased the power from the steam engine but also decreased fuel consumption. The United States was initially slower than England in adopting high pressure boilers, Table 1.2. Increased pressure in the boiler meant a larger flue gas exit temperature.

Process efficiency could be increased by decreasing the flue gas exit temperature. Installing economizers in the flue gas canals behind the boilers was one means of doing this (Barth, 1911). (The incoming water was heated in the economizers.)

The problem remained of how to build boilers that could withstand increasing pressures. The next innovative step was the emergence of the drum boiler. This coincided with the spread of a new manufacturing technology: forming. This allowed a cheap and reliable joint to be made between a drum and a tube. This meant that boilers could be built with just

Table 1.2 Increase of working pressure (MPa) of boilers in England and the United States

Year	1800	1818	1830	1843	1852	1870	1888	1900
England	0.12	0.26	0.34	0.40	0.66	0.83	1.00	1.37
United States	0.14	0.15	0.17	0.19	0.22	0.51	1.25	2.07

Source: Data from Hills, 1989 and Thurston, R.H., 1897. Promise and potency of high pressure steam. Transactions of ASME, vol. 18. pp. 1896—1897.

tubes and drums. A new class of boiler—the multiple drum tube boiler—had been born. Some early designs incorporated a number of drums (Stultz and Kitto, 1992). Soon a boiler with at least two drums became standard.

The larger the diameter and the higher the operating pressure, the thicker the shell needed to be. Therefore the size of the shell boiler was limited to low pressures and small sizes. If larger units were required, multiple boilers needed to be operated. In late 1800 some 10 water tube boilers could be connected to a single steam engine or turbine. With drums and tubes connecting them, much larger boilers than before could be built. The steam drum was beneficial and the manufacture of larger units was much easier. With this design there was better control of water quality. A mud drum could be used to blow out unwanted contaminants.

The requirement for still greater unit sizes meant larger furnace volumes were needed. This meant that combustion and heat transfer were separated. Previously, refractory-lined furnace walls were lined with water tubes (Effenberger, 2000). The first boilers with water tubes on all walls were introduced in the 1920s.

The late 19th and early 20th centuries witnessed the emergence of electric lighting for individual homes and electric utility business. Pioneering work was done by Thomas Edison in the United States. This meant that the steam generation market started to grow fast and required the unit size to increase rapidly (Thomas, 1975). Steam engines were soon replaced by steam turbines, and the steam generation capacity of a single boiler climbed steadily upwards.

The early boilers did not have superheating as steam engine efficiency did not improve significantly with the inlet steam temperature. It was not until after the introduction of the steam turbine by Laval that superheaters started to be built. Reheating was first used in the 1930s but remained scarce until after the World War II.

1.1.1.4 Forced Flow Boilers

Economics drove the engineers to continue to increase the efficiency and unit size of boilers. To further increase efficiency the maximum allowable working pressure limit by natural circulation needed to be replaced. Two solutions emerged. In 1928 the La Mont forced circulation boiler was patented. In it the natural circulation was aided by a pump. Another solution involved replacing the steam-water side configuration with a straight flow path. The first supercritical boilers emerged c1930. Depending

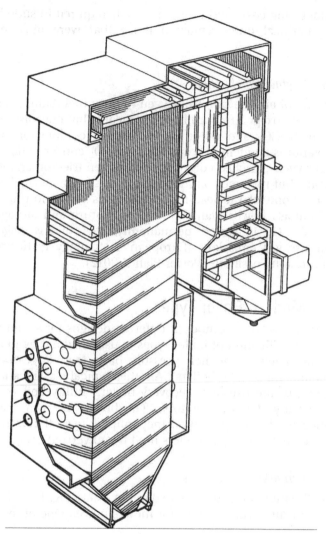

Figure 1.6 Benson boiler (Stultz and Kitto, 1992).

on the layout of their surfaces and the owner of the patent, they were called Sulzer, Benson, and Ramzin boilers (Fig. 1.6).

One of the early pioneers of once-through boilers was Englishman Mark Benson. His first idea for a once-through boiler was registered in 1923. It was not until 1957 that the first commercial boiler operated above 2212 MPa pressure. This unit at Ohio Power Company's Philo plant had an output of 85 kg/s steam at 31.37 MPa and 621°C. Because of the high manufacturing costs and exotic materials required in superheaters, subcritical units remained those that were most often built.

1.1.1.5 Oxyfuel Combustion

Instead of air one can use oxygen to burn fuel (Stanger et al., 2015). Using oxygen to burn biomass means that flue gases contain mostly CO_2 and water vapor. Oxyfuel combustion when water vapor is condensed results in high CO_2 concentration flue gases. Oxygen-enriched combustion has been tried on an industrial scale, but no large commercial unit exists yet. Oxyfuel combustion is one of the leading technologies for capturing CO_2 from biomass power plants with CO_2 capture and storage. If produced CO_2 is stored permanently, this means negative greenhouse gas emissions (European Technology Platform for Zero Emission Fossil Fuel Power Plants, 2012).

1.1.2 Modern Boiler Types

There are a large number of different designs for steam generators. Classification of boiler types is usually done by looking at the arrangement of their pressure parts (e.g., large-volume boilers and water tube boilers). Another typical classification is according to the use (e.g., recovery boilers, utility boilers, and heat recovery steam generators). The boilers are can also be classified by the firing method they employ (e.g., fluidized bed boilers, grate boilers, and boilers fired by pulverized coal).

1.1.2.1 Large-Volume Boilers

The fire tube boiler was created in the 1800s, Fig. 1.7. Its main use was to run steam engines for motive power. One of the first

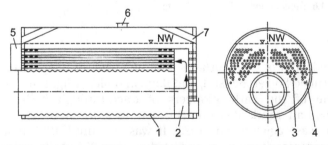

Figure 1.7 Typical design of industrial fire tube boiler: 1—Fire tube; 2—Smoke chamber; 3 and 4—Gas tubes; 5—Gas exhaust; 6—Steam outlet (Effenberger, 2000).

brand names for this type of boiler was the Lancashire boiler, which used two parallel fire tubes. It was used to power steamboats and railroad engines and to run industrial machinery via belt drives. The main design features of fire tube boilers are pressure $1-1.6-2.4$ MPa, max. 4.0 MPa, steam temperature saturated, max. 400°C, and steaming rate ~ 10 kg/s, max. 50 kg/s. In addition to steam generation, a large number of fire tube boilers are used to heat hot water for residential or commercial use.

Fire tube boilers are large-volume boilers. A large vessel, filled with water, is pierced with tubes of various sizes. Radiative heat transfer occurs mainly in the fire tube at the bottom. Convective heat transfer takes place in the large number of parallel gas tubes in the top. Hot flue gas transfers heat through tubes to the water. Saturated steam is led out from the top. The number of flue gas passes depends on the unit size and ranges from 2 to 4.

1.1.2.2 Pulverized Solid Fuel Fired Boilers

Most thermal power plants are built to fire pulverized solid fuels, especially coal. Pulverized coal-fired units, which can cofire biomass, tend to be large. They often have a tower construction. This means that all heat transfer surfaces are placed on top of each other, with no backpass. The benefits of coal-fired units are reliability and proven technology. The drawback is their fuel. Coal generates a large amount of CO_2. Reducing nitrogen oxide (NOx) and sulfur dioxide (SO_2) requires large additional investments in modern plants.

Pulverized firing can also be applied to biomass fuels. The drawback is the large amount of electric efficiency that is required to crush biomass to the required, small diameter particles. Pulverized firing has successfully been applied to peat. Very recently several large ($10-100$ MWth) boilers firing pellets have been built.

1.1.2.3 Fluidized Bed Boilers

The 1970s saw the introduction of a new combustion method: fluidized bed firing (Caillat and Vakkilainen, 2013). Fluidized bed firing development started in 1922 when Winkler patented his coal gasification concept. The first use of a circulating fluidized bed (CFB) as a chemical reactor was patented by Lurgi in the 1960s. In fluidized bed firing a hot inert bed is used to maintain stable conditions. Fluidization means that the inert bed is mixed constantly and well. The stable temperature promotes fast combustion and, with a low furnace temperature,

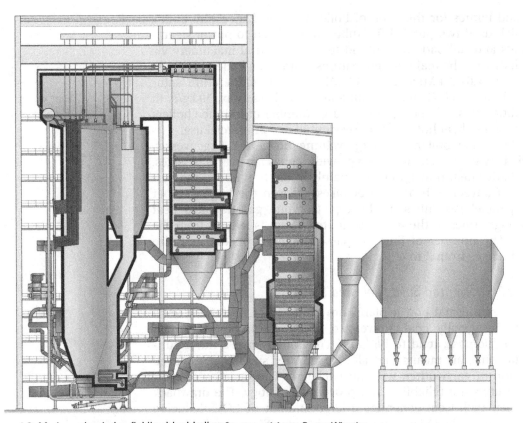

Figure 1.8 Modern circulating fluidized bed boiler. Courtesy of Amec Foster Wheeler.

excellent emissions. Only a small percentage of the fluidized bed boiler bed mass is from the burning fuel.

There are two main types of fluidized bed boilers: the bubbling fluidized bed (BFB) boiler and the CFB boiler. In the BFB boiler the bottom of the furnace has a solid/gas suspension, known as a fluidized bed, which behaves like a liquid. In the CFB boiler, Fig. 1.8, there is no sharp decrease of solid particles but solids travel continuously from the dense bottom of the furnace to the top. The CFB furnace can be said to consist of a suspension of sand, ash, flue gas, and burning fuel at a fairly even temperature. The solids drawn from the furnace with the CFB flue gases have to be separated and recirculated back to the lower furnace.

The main application of the BFB is in units smaller than 300 MWth. The CFB is used in larger units up to a size of +500 MWe (Hupa, 2005), and even larger units of 800 MWe have been proposed (Giglio, 2012). Fluidized bed firing is used especially with biofuels and when emission performance requires low NOx or low sulfur emissions. Low NOx is achieved because of the low furnace temperature (low bed temperature). SO_2 control is easily achieved because calcium carbonate can be inserted into the bed, where it reacts with the sulfur. The main reason for the lack of use of BFB technology in large units is that the bed itself must be made very large (Singer, 1991).

BFB firing is very suited to retrofitting old boilers. There have been many cases where old grate boilers have been retrofitted by replacing the lower part of the furnace.

1.1.2.4   Heat Recovery Boilers

Heat recovery steam generators (HRSGs) are very popular. They are often combined with a gas turbine or diesel generator, and produce additional electricity with the steam. These combined cycle units have very high electricity-generating efficiencies and good partial load characteristics. There have been several studies suggesting that biomass could be used in these combined cycle plants, but commercially they are rarely used.

There are two main arrangements. In a horizontal tube HRSG the evaporative tubes are placed horizontally. This type of arrangement is very popular in large units. In horizontal tube HRSGs the flue gas passage is often square. Vertical-type HRSGs have their main heat exchanger tubes in a vertical position. HRSGs of this type are popular with smaller unit sizes and are often much higher than they are wide (Fig. 1.9).

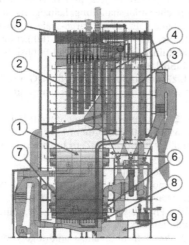

1 - Furnace
2 - Superheater
3 - Boiler bank
4 - Economizer
5 - Steam drum
6 - Air ports
7 - Liquor guns
8 - Smelt spouts
9 - Dissolving tank

Figure 1.9 Recovery boiler. courtesy of Andritz.

1.1.2.5 Recovery Boilers

In a recovery boiler concentrated black liquor burns in the furnace. Simultaneously inorganic process chemicals are reduced to be subsequently used in kraft pulping. Thus the recovery boiler serves three main purposes (Vakkilainen, 2005): First, it burns the organic material in the pulping residue (black liquor) to generate high pressure steam. Second, it recycles and regenerates used chemicals in the black liquor. Third, it minimizes discharges from several waste streams in an environmentally friendly way.

1.1.2.6 Package Boilers

The name package boiler comes from the fact that these boilers can be brought to the site as a single piece. Their transportation resembles the handling of a large package or container. The whole boiler is manufactured at the boiler shop. Package boilers are used in small plants to fire biomass. Their steam pressure is less than 6.4 MPa. Because of their compact structure, the investment cost is low and the delivery time is short. Steam drum, mud drum, and wall placement differ in the main types of package boiler.

References

Ångteknik, 1945. Aktiebolag De Lavals Ångturbin, Stockholm, 190 p.

Barth, F., 1911. In: German, G.J. (Ed.), Die dampfkessel, band II, bau und betrieb der dampfkessel. (Steam boilers, part II, construction and use of steam boilers). Göschen'sche verlagshandlung, Leipzig, 160 p.

Caillat, S., Vakkilainen, E., 2013. Large-scale biomass combustion plants: an overview (Chapter 8). In: Rosendahl, L. (Ed.), Biomass Combustion Science, Technology and Engineering. Woodhead Publishing Series in Energy, London, pp. 274–296. ISBN 9780857091314.

Croft, T. (Ed.), 1922. Steam boilers. first ed. McGraw-Hill, New York.

Dickinson, H.W., Jenkins, R., 1981. James Watt and the steam engine. Moorland Publishing, p. 415. ISBN 0903485923.

Effenberger, H., 2000, Dampferzeugung. (Steam boilers), Springer Verlag, Berlin, 852 p. ISBN 3540641750 (In German).

European Technology Platform for Zero Emission Fossil Fuel Power Plants, 2012. Biomass with CO_2 Capture and Storage (Bio-CCS); The way forward for Europe. Joint Taskforce Bio-CCS, June 2012, 32 p.

Forsman, W.V., Saraoja, E., 1928. Höyrykoneoppi (Steam machinery theory). Fifth Printing, Otava, Helsinki. 205 p. (In Finnish).

Giglio, R., 2012. CFB set to challenge PC for utility-scale USC installations. Power Eng. Int. January, 16–21.

Hills, R.L., 1989. Power from steam. Cambridge University press, p. 338. ISBN 0521343569.

Hunter, L.C., 1985. A history of industrial power in the United States 1780–1930, vol. 2: steam power. University Press of Virginia, Charlotteville, p. 732. ISBN 0813907829.

Hupa, M., 2005. Fluidized bed combustion. Politecnico di Milano Lecture series, Options for Energy Recovery from Municipal Solid Waste, January 31st–February 2nd, 2005, 41 p.

Intergovernmental Panel on Climate Change (IPCC), 2011. Special report on renewable energy sources and climate change mitigation: Chapter 2 Bioenergy. Intergovernmetal Panel on Climate Change, Working Group III—Mitigation of Climate Change, 180 p.

International Energy Agency, 2014. World Energy Outlook 2014. International Energy Agency, Paris, France, 748 p. ISBN 9789264208056.

International Energy Agency, 2015. Medium-Term Coal Market Report 2015: Market Analysis and Forecasts to 2020. International Energy Agency, Paris, France, December 2015, 166 p. ISBN 9789264249950.

Ishigai, S., 1999. Historical development of straregy for steam power. In: Ishigai, S. (Ed.), Steam Power Engineering. Cambridge University Press, 394 p. ISBN 0521626358.

Lindqvist, S., 1984. Technology on trial: the introduction of steam power technology into Sweden, 1715–1736. Ph.D. Thesis, Uppsala studies in history of science 1, Stockholm, Almqvist & Wiksell. 392 p. ISBN 9122007164.

Lobben, P., 1930. Handboken för mekaniker (Mechanic's handbook). fourth ed. Albert Bonniers Förlag, Stockholm, 748 p. (in Swedish).

Morris, I., 2010. Why the West Rules—for Now: the Patterns of History, and What They Reveal About the Future. Farrar Straus Giroux, New York, 750 p. ISBN 9780374290023.

Rogan, J.B., 1996. Comparison of current technology options for power generation in North America. In: Global Climate Change Forum, Public Utilities Commission of Ohio, September 16–17, 1996.

Shields, C.D., 1961. Boilers, Types, Characteristics, and Functions. McGraw-Hill, 559 p. ISBN 070568014.

Singer, J.G. (Ed.), 1991. Combustion Fossil Power. fourth ed. Asea Brown Boveri, 977 p. ISBN 0960597409.

Stanger, R.J., Wall, T., Spörl, R., Paneru, M., Grathwohl, S., Weidmann, M., et al., 2015. Oxyfuel combustion for CO_2 capture in power plants. Int. J. Greenhouse Gas Control. 40, 55–125.

Stultz, S.C., Kitto, J.B. (Eds.), 1992. Steam Its Generation and Use. 40th ed. 929 p. ISBN 0963457004.

Thomas, H.-J., 1975. Thermishe Kraftanlagen (Thermal Powerplants). 386 p., ISBN 3540067795. (in German).

Thurston, R.H., 1897. Promise and potency of high pressure steam. Transactions of ASME. 18, 1896–1897.

Vakkilainen, E.K., 2005. Kraft recovery boilers – Principles and practice. In: Suomen Soodakattilayhdistys, R.Y. (Ed.), Valopaino Oy, Helsinki, Finland, 246 p. ISBN 9529186037. Available from: <http://www.doria.fi/handle/10024/111915>.

2

SOLID BIOFUELS AND COMBUSTION

Steam Generation from Biomass. DOI: http://dx.doi.org/10.1016/B978-0-12-804389-9.00002-2

One of the main uses for wood has always been to produce energy. Cheap oil, especially after the Second World War, meant that the use of wood for energy generation declined. Currently the use of biofuels is increasing but there are only a few countries in the world where biofuels have a significant share in energy use. In Europe some countries such as Finland, Sweden, Austria, Portugal, and France use wood in significant amounts to meet the energy needs. Less than half of the biomass in the European Union (EU) is used for energy (Mantau et al., 2010). Energy policies call for energy security, increased renewable energy use, and limiting climate change. This encourages the use of biomass as a source of energy. In addition, high fossil fuel (oil and coal) prices since 2000 have meant that bioenergy is again in fashion.

In addition to wood and short rotation coppice crops (SRC), examples of renewable, sustainable biomass fuel sources are thinnings and various agricultural residues such as sugarcane waste (bagasse), timbermill waste (sawdust), forestry residues, straw, palm oil, rice husks, coffee husks, nut shells, coconut residue, and olive processing by-products (Arbon, 2002).

It is estimated that there are about 40 million square kilometers or 420×10^9 tons of forest worldwide (Parikka, 2004). This is roughly a third of the earth's land area. About half of the world's forests are tropical and subtropical. The next biggest forest mass is the temperate and boreal forests in the Northern hemisphere. A very large fraction of globally harvested wood is used as a renewable energy source. Biomass currently represents a little less than 10% of the world's primary energy consumption (IEA, 2014). About one-third of the biomass usage for primary energy is in the industrialized countries. Biomass is used for generating electricity and heat. A large fraction of this comes from residues of biomass-based industries used for process heat production. This is called modern usage. In modern usage, efficiency is high and environmental protection has been accounted for. Two-thirds of biomass usage is labeled traditional usage. Various kinds of biomass are used in developing countries, typically for cooking and household heating needs.

Solid biomass can mitigate climatic change caused by greenhouse gas emissions when it is substituting fossil fuels. Increased use of sustainably produced, renewable biomass could achieve a significant reduction to global CO_2 reduction especially by producing various types of biofuels to replace fossil fuels (WWF, 2011).

The EU has passed the RES Electricity Directive (2001) to increase the share. In it the biomass has been defined to mean

the biodegradable fraction of products, waste and residues from agriculture (including vegetal and animal substances), forestry and related industries, as well as the biodegradable fraction of industrial and municipal waste

Biomass is a material of biological origin. This includes both plant- and animal-derived materials. Various industries, such as pulp and paper, agricultural, and the food industry produce biomass as residue. This biomass is often utilized for energy, but also partly wasted. Biomass can also be specifically grown for energy use. Examples of different biomasses used for conversion and energy production around the world are:

- wood, demolition wood, wood residue
- black liquor
- waste
- short rotation crops, grass
- corn
- straw
- manure
- organic waste
- jatropha
- microalgae.

Biomasses derived from wood currently account for about 80% of biomass feedstocks that are used for energy (Chum et al. 2011) (IPPC, 2011). The rest are from the agricultural sector and from various waste and by-product streams. In industrial countries the forests are mainly used sustainably. In modern usage biomass has a price. So looking after your revenue makes sense. In many parts of the world traditional biomass usage, especially fuelwood logging, exceeds the regeneration capacity of the forests. This is mainly because the biomass is considered free and without monetary value. Bioenergy feedstock production often competes with other uses of land and water. Efficient land use, especially in developing areas, secures food and feed supply and provides rural workplaces through increasing biomass production (IEA, 2014).

Different combustion techniques make different demands on the biofuels' characteristics, which is important if problems are to be avoided or minimized. Table 2.1 gives a summary of the biofuels' characteristics that are important for different combustion techniques. The table is intended to serve as a guide for the requirements that should be made on the biofuels' properties in a particular combustion situation.

Table 2.1 Important Biofuels' Characteristics for Different Combustion Installations

Characteristics	Pulverized	Grate	Fluidized Bed
Design fuel	Designed for specific fuel	Designed for specific fuel	Able to fire several
Heating value	Typically over 15 MJ/kg	From 5 MJ/kg	Wide
Moisture content	Almost dry	Constant range 30–55%	Dry – 70%
Ash content and quality	High ash means larger furnace	Requires low ash content	Up to 50%, sensitive to low melting point ash
Volatiles content	Insensitive	Insensitive but designed for certain range	Insensitive but requires enough fuel insertion points
Particle size	Lower than 1 mm, fines burn high	Higher than 10 mm, sensitive to fines, sticks, strips, and large particles	1–100 mm, furnace operation insensitive

Source: Adapted from Strömberg, B., 2006. Fuel handbook. Technical Report, SVF/971 Stiftelsen för Värmeteknisk Forskning, Stockholm, Sweden, 105 p. (Strömberg, 2006).

2.1 Classification of Solid Biofuels

Solid biomass fuels have recently (less than 500 years ago) been formed by photosynthesis. They are classified as renewable fuels. The other main class of solid fuels, fossil fuels, used to be biomass fuels, but have undergone slow reactions underground taking place from thousands to millions of years. Because they cannot be reproduced within a reasonable time, they are classified as non-renewables. Combustion of renewable solid biomass does not increase the carbon dioxide (CO_2) in the global atmosphere. Carbon dioxide released during the burning of renewable fuels can be assumed to be recycled to the carbon dioxide that forests use for growing. Biofuels are highly volatile. This can be associated with their relatively high oxygen (O) and hydrogen (H) content. The portion of the volatiles in most biofuels is around 70% to 80%. The carbon content and heating values are low compared to fossil fuels (Caillat and Vakkilainen, 2013).

Table 2.2 shows examples of biomasses and their properties. The properties of wood vary according to the wood species, origin, and age. Therefore there is a wide variation in the properties of wood-based biomasses. Especially the properties

Table 2.2 Examples of Biomass Feedstocks and Their Properties

Biomass	Ash Content of Dry Matter %	Moisture %	Lower Heating Value (Dry Matter) MJ/kg
Wood chips	0.5–2	40–55	19.0–20.5
Wood shavings	0.4–0.5	5–15	18.9–19.2
Sawdust	0.4–1.1	45–60	18.9–19.2
Forest residue	1–3	50–60	19.3–20.1
Bark	1–3	45–65	18.6–2.7
Straw	4.5–6.5	17–25	16.7–17.8
Reed canary grass	5.9	14.4	17.7
Manure (cow)	16	14	19.2
Animal fat	0.1	0.4	36.5
Vegetable fat	0.1	0.6	36.9
Olive residue	4.5	8.1	20.3

Source: LUT database; Alakangas, E., 2000. Suomessa käytettävien polttoaineiden ominaisuuksia. (Properties of Finnish fuels). VTT Research Notes, 2045 VTT Energia, Espoo. 189 p. ISBN 9513856992 (in Finnish) (Alakangas, 2000).

of straw and other fast-growing biomasses depend on the plant species, weather, and soil condition during their growth.

Biomass by definition is biological material from living, or recently living, organisms. Often biomass is used to mean plants, especially wood. Recently biomasses from animal-, vegetable-, and fishery-derived material have been used for energy (Williams 1992; Jenkins et al., 1998). A recent classification of biofuels is the European standard for solid biofuels (CEN TC 335). Biomass fuels can generally be divided into four primary classes based on their source (Khan, 2007):

- wood and woody fuels (hardwood, softwood, demolition wood)
- herbaceous fuels (straw, grasses, stalks, etc.)
- waste (sewage sludge, refuse-derived fuel (RDF) waste from paper and food industries, etc.) and
- energy crops (specifically cultivated for energy purposes).

2.2 Biomass Conversion to Energy Uses

The main modern energy use for solid biomass is the direct production of heat and power. This remains the most important use for biomass. Refining to improve biomass fuels is on the

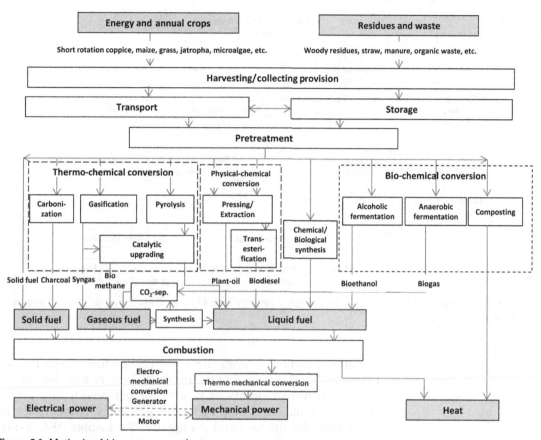

Figure 2.1 Methods of biomass conversion to energy. After the IPCC Biomass Report, draft 2010. (Chum, 2012)

rise due to growing global interest in, for example, transport biofuels (Vakkilainen et al., 2013). A large variety of biomasses is used for energy. The economics and yields of biomass feedstock vary widely between different biomass types and also regionally (Chum et al., 2011). Economics, local availability, and the maturity of technology mean that wood, process residues, and bio-waste are primarily used for heat and power production. Agricultural crops are the main source for renewable liquid biofuels (Fig. 2.1).

Biomass conversion methods for energy purposes are shown in Fig. 2.2. Thermochemical methods require heat usage in the process. In biochemical conversion the biomass molecules are broken down into smaller molecules using bacteria or enzymes. These processes require only little, if any, external energy. Physical chemical conversion is mostly done by various machines using mechanical energy.

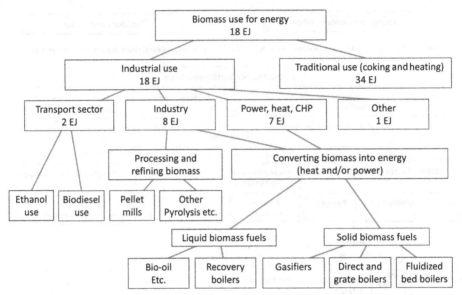

Figure 2.2 Categorization of biomass users for the IEA Bioenergy Task 40 study: Large industrial users of energy biomass. (Vakkilainen et al., 2013)

Large conversion units with increasing fuel flexibility are often preferred for economic reasons. This is true for steam generation from biomass as well as biofuel production. Lower specific capital costs and increased efficiency compensate in many cases for the cost and energy use of transport (IEA, 2014). The handling and transport of biomass forms about 20% to 50% of the total costs of bioenergy production. One of the reasons for the increase in solid biomass usage has been the decreased cost of solid biomass supply chains by even more than 50% (Chum et al., 2011). Scale increases, technological innovations, and increasing competition have led to the declining price of harvesting, collection, and transport.

Biofuels and biomass are little traded, compared to bioenergy products, but trade flows are growing rapidly. Fuelwood is regarded as a local product and very little of the production is traded. Sawdust and wood waste for energy purposes are mostly traded in the form of pellets. Wood chips for energy purposes are not transported such long distances as wood pellets. Wood chips are mainly traded for purposes other than energy—primarily pulp and paper production—and only 10% of the trade is estimated to be energy-related. Wood chip trade for energy purposes takes place mainly to and within the EU (Lamers et al., 2012).

There are various estimates on the future potential of biomass for energy use. Chum et al. (2011) estimate that technical potential of biomass for energy might be as much as 500 EJ per annum by 2050. The International Energy Agency (IEA, 2014) estimates that total biomass energy potential, up to 2050, will be between 40 and 1100 EJ. The higher value requires intensive agriculture, which concentrates on soil of good quality, while the lower value scenario assumes that only residue can be utilized. The report estimates, e.g., forest residue potential to be 30−150 EJ, agricultural residue potential 15−70 EJ, and biomass production potential on marginal lands from less than 60 EJ up to 110 EJ.

2.3   Biomass Usage for Energy

The largest part of biomass usage is traditionally for cooking and heating purposes in developing countries (IEA Statistics, 2012). Industrial use in industrialized countries accounts for about a third of the total biomass use for energy. Industrialized countries use biomass for power and heat generation, transportation, and other purposes, which are mainly heating and cooking. The categorization of biomass users is represented in Fig. 2.2.

The largest users of biomass can be found in the transportation sector. The largest plants are ethanol mills and biodiesel plants. They refine biomass for motor fuel use. There are also very large plants that convert biomass into heat and power. A few pellet mills are large, but existing torrefaction and pyrolysis plants are scarce and small. The largest sector of heat and power generation plants that use biofuels are black liquor recovery boilers. Plants that use solid biomass fuels; such as fluidized bed boilers, pulverized fuel-fired boilers, grate boilers, and gasifiers, are of various sizes.

2.3.1   Biomass Use at Country Level

Total usage of biofuels and waste for energy purposes was 51.8 EJ in 2009 (IEA Statistics, 2012; Kuparinen et al. 2014). Fig. 2.3 shows the distribution of biomass consumption between different types of user. Residential cooking and heating, mainly in developing countries, use most of the biomass—34 EJ. use of biomass had a share of 15%, which equals 7.8 EJ. Only 2.3 EJ (4%) was used for electricity generation. This is roughly equal to the transportation sector, which used 2.2 EJ (4%). Combined heat and power (CHP) plants used 1.3 EJ (3%).

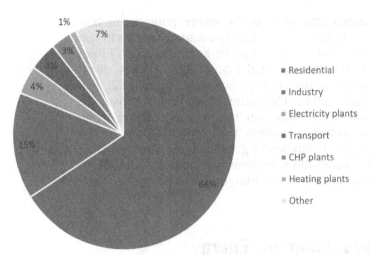

Figure 2.3 Biomass use for energy purposes in 2009. Data from International Energy Agency, IEA energy statistics, 2012. Available from: <http://www.iea.org/stats/index.asp> (Accessed 22.05.12).

The top 20 countries that used the most biomass for energy purposes in the world in 2009 are listed in Table 2.3. In the table it can also be seen how biomass usage is distributed between power and heat production, industrial use, transportation use, and other sectors. The figures in the table include liquid, gaseous, and solid biomass, but not waste usage. Energy industry use includes both electricity and heat production and the energy industry's own use. Other use is mainly residential use and use in commercial and public services and agriculture.

Table 2.4 shows the 20 countries that use the most biomass for energy production. Waste use is not included. The figures include heat and electricity production and the own energy use.

There are also countries where biomass is not used at all or only in very little amounts. IEA Statistics include the data of energy use of 136 countries, of which 17 did not use biomass for energy purposes, or used amounts that are not large enough to show. These countries are mainly relatively small, and most of them are either island states, such as Iceland and Malta, or located in the Western Asia region, such as Kuwait, Qatar, and Saudi Arabia.

2.3.2 Use of Biomass in Industrial Applications

Biomass usage in the industrial sector by country in 2009 is listed in Table 2.5. The 20 countries listed used 6.1 EJ of solid

Table 2.3 Biomass Use for Energy Purposes in the Top 20 Biomass-Using Countries in 2009 (The Use of Waste Is Excluded)

No.	Country	Biomass Use PJ	Industry PJ	Transport PJ	Energy PJ	Other PJ	Of Total %
1	China	8525	0	52	54	8420	17
2	India	6926	1195	7	48	5676	14
3	Nigeria	3848	379	0	102	3367	8
4	Brazil	3178	1317	551	857	447	6
5	United States	3089	1170	928	471	519	6
6	Indonesia	2202	272	1	30	1898	4
7	Ethiopia	1258	0	0	70	1188	3
8	Pakistan	1236	134	0	25	1077	2
9	Vietnam	1053	0	0	36	1017	2
10	Congo	899	185	0	20	694	2
11	Thailand	863	283	22	306	252	2
12	Germany	807	95	116	385	209	2
13	Tanzania	720	79	0	89	553	1
14	Kenya	596	0	0	212	384	1
15	South Africa	593	75	0	158	359	1
16	France	535	98	103	37	297	1
17	Canada	467	287	33	62	84	1
18	Sudan	450	33	0	177	240	1
19	Myanmar	441	13	0	16	411	1
20	Sweden	389	169	16	160	44	1
	Other countries	9453	1461	329	1979	5706	17
	World total	50,197	7504	2158	5340	35,195	100

Source: Calculated from International Energy Agency, IEA energy statistics, 2012. Available from: <http://www.iea.org/stats/index.asp> (Accessed 22.05.12).

biomass a year. This is 83% of the world total solid biomass usage in the industrial sector (IEA Statistics, 2012). The total use was over 7 EJ a year. Industrial use was mostly primary solid biomass. In Germany and the United States, biogases and bioliquids were used. Solid biomass usage dominated in other countries.

As the largest biomass-based energy user country in the industrial sector, India used 1195 PJ of biomass in 2009. The high figure can be explained by the fact that, in addition to

Table 2.4 The Largest Biomass-Using Countries in the Energy Sector, Including Heat and Power Production and the Own Energy Use

No.	Country	Use for Energy Production PJ	Of World Total %
1	Brazil	857	16.1
2	United States	471	8.8
3	Germany	385	7.2
4	Thailand	306	5.7
5	Kenya	212	4.0
6	Sudan	177	3.3
7	Sweden	160	3.0
8	South Africa	158	3.0
9	Côte d'Ivoire	154	2.9
10	Japan	115	2.2
11	United Kingdom	111	2.1
12	Nigeria	102	1.9
13	Finland	100	1.9
14	Tanzania	89	1.7
15	Angola	78	1.5
16	Mozambique	75	1.4
17	Italy	74	1.4
18	Ethiopia	70	1.3
19	Ghana	66	1.2
20	Australia	66	1.2
	Other countries	1467	27
	World total	5340	100

Calculated from International Energy Agency, IEA energy statistics, 2012. Available from: <http://www.iea.org/stats/index.asp> (Accessed 22.05.12).

wood, India uses annual crops such as bagasse, rice husk, straw, cotton stalks, and coconut shells for power generation. If we look at the countries in the list, we note the influence of pulp and paper manufacture in one of them. It uses a significant amount of biomass to make wood-based products and burns the excess biomass. Another example is sugar production. Around the world, residue from sugar making, bagasse, is used for power generation. Biomass usage could be significantly increased if we looked at residual biomass in breweries, textile

Table 2.5 The Largest Biomass User Countries in the Industrial Sector

No.	Country	Liquids	Biogases	Primary Solid Biomass	Total	Of Total
		PJ	PJ	PJ	PJ	%
1	India	0	0	1195	1195	16.2
2	Brazil	0	0	1179	1179	16.0
3	United States	4	104	1063	1171	15.9
4	Nigeria	0	0	379	379	5.1
5	Canada	0	0	287	287	3.9
6	Thailand	0	0	283	283	3.8
7	Indonesia	0	0	272	272	3.7
8	Democratic Republic of Congo	0	0	185	185	2.5
9	Sweden	0	0	169	169	2.3
10	Pakistan	0	0	135	135	1.8
11	Finland	0	1	111	112	1.5
12	Australia	0	0	105	105	1.4
13	France	0	1	97	98	1.3
14	Japan	0	0	89	89	1.2
15	Tanzania	0	0	79	79	1.1
16	South Africa	0	0	76	76	1.0
17	Germany	16	13	71	100	1.4
18	Colombia	0	0	69	69	0.9
19	Spain	0	1	65	65	0.9
20	Sri Lanka	0	0	64	64	0.9
	Other countries	2	4	1246	1252	17.0
	World total	22	124	7220	7366	100

Source: International Energy Agency, IEA energy statistics, 2012. Available from: <http://www.iea.org/stats/index.asp> (Accessed 22.05.12).

mills, fertilizer plants, solvent extraction units, rice mills, and petrochemical plants (Aradhey, 2012).

As an example, the industrial sector in Brazil used 3587 PJ of energy in 2010. Sugarcane bagasse was responsible for 20.8% of that energy and 7.1% was generated from other renewable primary sources including pulp and paper. The majority of energy utilization of sugarcane bagasse is in the foods and beverage sector, where it covers 75% of the sector energy demand (MME, 2011). Wood residues are consumed for energy purposes mainly in the pulp and paper industry, while firewood is used mainly in the food and beverage sector and in the ceramic industry (Walter and Dolzan, 2012).

Most of the biomass consumed by the industrial sector in the United States is woody biomass. The largest source is black liquor. A third of the total industrial biomass energy consumption is wood and wood waste in the wood processing industry (Hess et al., 2010). In addition the United States has started to export pellets, which do not show up as energy use.

2.3.3 Use of Biomass in the Transportation Sector

The transportation sector no longer uses significant amounts of solid biomass. Liquid biofuels almost entirely make up the renewable portion. The 20 countries that used the most biomass in the transportation sector in 2009 are listed in Table 2.6. The world total usage of liquid biofuels for transportation in 2009 was 74 Mt amounting to 2.1 PJ. Of these 20 countries only Sweden used some biogas (IEA Statistics, 2012).

In 2009 ethanol was the largest biofuel in the United States, where biomass-based fuels accounted for about 2.5% of the total transport fuel usage. Nearly half of the global use of liquid biofuels for transportation was in the United States. Biodiesel usage accounted for only 2.7% of the renewable energy consumption in the transportation sector (Hess et al., 2010).

In Brazil, according to MME (2011), in 2011 17.3% of total fuel consumption in the transportation sector was bioliquids. On average, 87% of ethanol produced in Brazil is used as fuel. The rest is used by industry, such as chemicals and cosmetics companies (Barros, et al., 2011).

In Europe in 2010 78% of the biofuel used was biodiesel. Only 21% was ethanol and ethyl tertiary butyl ether (ETBE), and 1% was other biofuels (Beurskens, 2011). One of the largest users is Germany, where in 2010 the share of biofuels was 5.8%. The most used biofuel in Germany is still biodiesel (Thrän et al., 2012).

Even though the biomass usage in India is high, biofuels are not used in the transport sector. As India is the world's fourth largest petroleum consumer, this has prompted the Indian government to promote the blending of ethanol derived from sugar molasses or juice with gasoline and the blending of biodiesel derived from inedible oils and oil waste with fossil diesel. India aims to derive biofuels from inedible feedstock grown on areas that are unsuitable for food or feed production (Aradhey, 2012).

Table 2.6 Biofuel Consumption in the Transportation Sector, 2009 — the 20 Largest User Countries

No.	Country	In Total PJ	Liquid Biofuels PJ	Gas PJ	Biodiesel PJ	Ethanol PJ	Of World Use %
1	United States	917	917	0	55	863	43.0
2	Brazil	531	531	0	38	494	24.9
3	Germany	124	124	0	107	17	5.8
4	France	102	102	0	85	17	4.8
5	China	52	52	0	11	42	2.5
6	Italy	49	49	0	44	4.9	2.3
7	Spain	45	45	0	38	6.1	2.1
8	United Kingdom	41	41	0	34	6.7	1.9
9	Canada	33	33	0	3.2	30	1.6
10	Poland	26	26	0	19	7.3	1.2
11	Thailand	23	23	0	16	7.6	1.1
12	Austria	21	21	0	19	2.6	1.0
13	Belarus	18	18	0	18	0.0	0.9
14	Sweden	16	15	0.9	6.8	8.1	0.7
15	Netherlands	16	16	0	10	5.6	0.7
16	Belgium	12	12	0	10	1.6	0.6
17	Australia	10	10	0	4.8	5.6	0.5
18	Portugal	9.3	9.3	0	9.3	0.0	0.4
19	Republic of Korea	9.0	9.0	0	9.0	0.0	0.4
20	Colombia	8.6	8.6	0	5.3	3.4	0.4
	Other countries	71	71	0	48	23	3.3
	World total	2131	2131	0.9	574	1557	100

2.3.4 Straw and Grass

There are often operating problems if conventionally designed boilers are used to burn large amounts of agricultural residues such as straw (Miles et al., 1995). Annual crops produce a high amount of ash. It is hard to remove external material clinging to biomass, such as soil. Annual crops typically have a high alkali content. Straw in particular forms very problematic ash deposits on heat transfer surfaces.

Limestone can be used to improve operation. Calcium appears as a constituent of deposits on convection surfaces (as $CaCO_3$, $CaSO_4$) and often reduces but does not prevent deposition. High alumina sand reduces agglomeration in a circulating fluidized bed (CFB) but does not change the composition of deposits on the superheater tubes.

2.4 Biomass Combustion

Biomass combustion can be thought to occur in four stages (Fig. 2.4, Table 2.7). In the first stage the biomass dries as fuel enters the furnace and it starts receiving heat. Heat increases

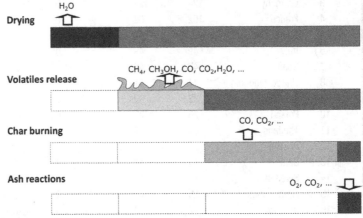

Figure 2.4 Processes occurring during biomass combustion.

Table 2.7 The Main Stages in Biomass Combustion

Stage	Characterized by	Time Scale in Furnace, Seconds for a 2 mm Piece
Drying	Water evaporation No or little coloring on fuel surface	0.05–0.2
Devolatilization	Appearance of flame, ignition Some swelling Release of volatiles Bright, radiating surface	0.1–0.3
Char burning	Disappearance of flame Decreasing diameter Reduction reactions Hot, yellowish-reddish surface	0.2–10
Ash	Constant or increasing diameter Reoxidation, carbonation	Long

the surface temperature in the biomass and starts evaporating water from it. Inside the fuel particle the temperature, and thus the partial pressure of water vapor, increases. Generated water vapor exits through the micropores in the fuel.

Volatiles release is the second stage, in which flames appear during the burning of firewood in the fireplace. When the fuel particle temperature increases, the small molecule weight compounds start to appear. Weak chemical bonds break and the small organic molecule population increases inside the fuel. The gas pressure increases and gases exit through the micropores in the burning fuel particle. The flame appears on the surface. Typically the reaction rate is high enough so that oxygen cannot penetrate inside the burning biomass particle.

After the volatiles release, the remaining char continues to have reactions. The fuel particle temperature increases and by this third stage most of the hydrogen has reacted. The main remaining combustible component is carbon. This remaining carbon reacts slowly through CO-gasification, H_2O-gasification and to a lesser extent with oxygen reactions. Flue gases exit through the micropores in the char and the fuel particle glows.

In the fourth stage the remaining ash undergoes further reactions, depending on the surroundings. Most of the biomass has reacted and only the inorganic part remains (Fe, SO_4, CO_3, Mg, Si, Ca, etc.). We often think of the ash as stable and unreactive but the remaining inorganics continue their reactions (e.g., oxidation). The ash particle formation continues with possible agglomeration and surface deposition.

2.4.1 Drying

Drying is characterized by the evaporation of water from the biomass particle. Drying can be experimentally defined as the period up to the beginning of combustion reactions (the visible flame). The drying of the biomass particle requires heat to evaporate the water. It typically proceeds as fast as the heat is transferred to the particle. Even in the furnace temperatures, drying is limited by the heat transfer from the surroundings to the particle.

In a typical biomass boiler the biomass surface temperature increases to $130°-160°C$ during the first couple of milliseconds after its insertion into the furnace. As water is evaporated, starting from the surface, the wet core size decreases. Typically the biomass particle is not completely dry at the center when the volatiles release starts close to the surface. Initially the biomass can be thought to consist of water and pyrolysable material that remains after drying. Biomass mass can then be divided into

$$m_0 = m_p + m_w \tag{2.1}$$

where

m_0 is the initial particle mass, kg
m_w is the mass of water in the particle, kg
m_p is the mass of pyrolysable material in the particle, kg
The dry matter at the start of the volatiles release (visible flame) is

$$x_d = \frac{m_p}{m_p + m_w} \tag{2.2}$$

where

x_d is the particle dry matter content
The heat to particle is ($c_p = 2.5$ kJ/kg, $l = 2450$ kJ/kg)

$$\frac{\partial}{\partial t} Q_d = m_p c_p \frac{\partial}{\partial t} T_d + l \frac{\partial}{\partial t} m_w \tag{2.3}$$

where

$\frac{\partial}{\partial t} m_w$ is the loss as water vapor
Heat to particle is the sum of convective and radiative heat

$$Q_d = Q_c + Q_r \tag{2.4}$$

The particle area is

$$A_d = 4^* \pi D_d{}^2 \tag{2.5}$$

The convective heat flux from hot flue gases to the colder particle is typically small, in the order of 10 W/m^{2}K

$$\frac{\partial Q_c}{\partial t} = h_c A_d (T_g - T_d) \tag{2.6}$$

where

A_d particle surface area, m^2
T_g gas temperature, K
T_d particle temperature, K
For the modeling of convective heat transfer a model of Ranz and Marshal (1952) can be used

$$Nu = \frac{h_c D_d}{\lambda} = 2.0 + 0.6 Re_d^{\frac{1}{2}} Pr^{\frac{1}{3}} \tag{2.7}$$

where

D_d particle diameter, m
h_c convective heat transfer coefficient, W/m^{2}K

Re_d Reynolds number based on particle diameter and relative speed

Pr Prandtl number based on gas phase

The radiative heat flux ($\varepsilon = 0.8$)

$$\frac{\partial Q_r}{\partial t} = \varepsilon A_d \sigma (T_g{}^4 - T_d{}^4) \tag{2.8}$$

The basic radiation heat transfer equation for particle in gray gas is

$$\Phi_r = \varepsilon \sigma A_d (T_g{}^4 - T_d{}^4) \tag{2.9}$$

This type of equation is applicable to the real combustion processes in biomass boiler furnaces. If more precise figures are needed, then computational fluid dynamics (CFD) calculations can be performed (Hyppänen and Raiko, 2002).

2.4.2 Devolatilization

As biomass temperature increases, reactions with the lowest activation energies start taking place. During devolatilization biomass fuels release light gases such as carbon monoxide, carbon dioxide, methane, and hydrogen. In addition to gases, tar has been reported to be the main product of devolatilization. Devolatilization is characterized by the appearance of a visible flame when released volatile gases from the biomass particle burn with the surrounding oxygen.

During devolatilization of solid biomass the gas release is large. The release of the volatile fraction occurs whether or not there is oxygen present. Because the flow of reduced gases is high, practically no free oxygen can reach the particle surface. The conditions inside the particle resemble those of pyrolysis or heating in an inert atmosphere. Devolatilization is often incorrectly called pyrolysis.

Devolatilization of biomass tends to be a fast process and depends essentially on heat transfer to the particle. Increasing the surrounding temperature typically results in a higher pyrolysis yield in laboratory conditions. For small solid biomass particles, devolatilization can be modeled using uniform temperature assumption. For larger (>1 mm) particles, devolatilization tends to occur in a reactive shell that starts on the surface, while a core of colder, unpyrolyzed material remains within (Järvinen et al., 2002; Saastamoinen, 2007).

During the solid biomass devolatilization, the biomass particles lose mass and increase porosity. The majority of the carbon in the fuel is consumed. Typically, in industrial biomass combustion processes in large boilers, carbon conversion of above 90% can be reached.

2.4.3 Char Combustion

Char combustion is defined as the period after the volatiles release has finished. The combustible material remaining after the volatiles release is often called fixed carbon. Fixed carbon does not include inorganic ash. In laboratory tests the start of char combustion starts when the visible flame is extinguished. In practice, char combustion and devolatilization of biomass fuels in industrial boilers overlap considerably. The term organic combustion time (i.e., the sum of devolatilization and char combustion times) can be used to characterize biofuel combustion (Frederick and Hupa, 1993).

2.4.4 Ash Reactions

What remains after combustion is called ash. Depending on the fuel transport and treatment, a large portion of the biomass ash can be from contamination by soil that entered the furnace along with the biomass fuel. In particular, silica content is dependent on the level of soil within the fuel. Old fuels (coal, peat) typically produce larger amounts of ash-forming compounds. For young fuels (biomass) there are often large amounts of alkali in the ash. These can cause fly ash deposits in the flue gas side as well as fluidized bed sintering (Zevenhoven-Onderwater et al., 2000). Typically we try to predict the behavior of different ashes by using fuel analysis, especially the analysis of inorganic constituents and thermodynamic equilibrium calculations. Ratios of inorganic components can point out which elements matter. Equilibrium calculations tell us what components are likely to form.

2.5 Basics of Combustion Calculations, Heat and Mass Balances

During combustion a fuel reacts with air or oxygen, producing heat and flue gases. It can be defined as the complete, rapid exothermic oxidation reaction of a fuel. Elements in a solid biomass fuel combine with a sufficient amount of oxygen or air. The result is combustion products called fuel gases (Teir, 2004). Combustion produces heat (i.e., it is exothermic). Biofuels consist primarily of carbon, hydrogen, and oxygen. Most of the carbon is converted in the combustion process to carbon dioxide (CO_2) and most of the hydrogen to water vapor (H_2O).

In addition to the three large elements, biofuels contain sulfur, nitrogen, and ash. Sulfur reacts mostly with oxygen to form sulfur dioxide (SO_2). About a third of the nitrogen in the fuel forms nitrogen oxide (NOx). Ash is often considered to be inert and unreactive for combustion calculations. Compounds in ash can, however, oxidize, reduce, and react. The gaseous product of combustion reactions is called flue gas.

2.5.1 Air Ratio

Depending on the amount of oxygen available for combustion reactions, three areas exist: stoichiometric, complete, and incomplete combustion.

In perfect, theoretical, or stoichiometric combustion all combustible elements in the fuel react using the theoretical minimum amount of oxygen. Stoichiometric air is a useful concept as it describes how much air is required for burning a specified amount of fuel if all combustion reactions proceeded to their most oxidized state. Because of mixing, temperature, and time restrictions, practical combustion at stoichiometric conditions in a biomass boiler produces unreacted combustion products such as carbon monoxide (CO). Per definition, the air flow required for stoichiometric combustion divided by the air flow required for stoichiometric combustion equals unity.

In incomplete combustion, where the air ratio is below unity, there is not enough oxygen for all combustible elements to react fully. This results in the formation of carbon monoxide, hydrogen (H_2), methane (CH_4), other volatile compounds (VOCs), tar, soot, and smoke. In industrial gasification the air ratio is often at or below 0.8. In fluidized bed boilers the air ratio in the lower furnace is often below unity.

Complete combustion would be achieved when all the reactive elements in the fuel were burned. In practice this is not possible. One always finds some unburned carbon with ash and some CO in flue gas. In solid biomass combustion the target is to use almost full combustion using the minimal amount of air. Biomass is typically fired at air ratios of 1.15–1.3 (i.e., 115–130% of the oxygen needed for stoichiometric combustion).

2.5.2 Chemical Reactions

In basic combustion calculations we assume that each carbon (C) atom in the biofuel reacts with one oxygen molecule (O_2) producing one molecule of carbon dioxide (CO_2). The reaction can be written as:

$$C + O_2 \rightarrow CO_2 \qquad (2.10)$$

As counting each carbon atom in a fuel would be difficult, we use a new unit called mole. Mole is a large number and can be defined as the number of atoms in 0.012 kg or the most typical isotope of a carbon atom that has the relative atomic mass of 12. In nature there are always some isotopes with higher and lower atomic mass. Therefore the mole weight used for carbon is not exactly 12 but a number very close to it. The National Institute of Standards and Technology (NIST) chemistry webbook gives the carbon mole weight as 12.0107 g/mol (Chase 1998). Similarly, in ideal conditions we can assume that hydrogen (H) and sulfur (S) react with oxygen as follows:

$$S + O_2 \rightarrow SO_2 \qquad (2.11)$$

$$H_2 + O \rightarrow H_2O \qquad (2.12)$$

The oxygen in fuel is thought to provide some of the required oxygen. Similar to ash the nitrogen (N) in the fuel is in basic combustion calculations assumed not to react. The water in fuel is assumed to evaporate but not to react.

If the biofuel has the following composition, then:

Lower Heating Value (LHV)	20.0	MJ/kgdf
Carbon	52.19	mass-% (% in dry fuel)
Hydrogen	6.31	mass-%
Nitrogen	0.55	mass-%
Sulfur	0.01	mass-%
Oxygen	39.74	mass-%
Ash	1.20	mass-%
Moisture content	57.00	%

One needs to be careful when looking at fuel compositions. One finds very differing practices, depending on the source. The fuel composition can be given per wet fuel mass or as fired (af), per dry fuel (df) or even per dry ash free (daf) fuel.

The input flow of biomass dry matter based on 1000 g of dry fuel is then:

	Mass, g/kgdf	Mol/kgdf	End Product
Carbon	521.9	521.9/12.011 = 43.452	CO_2
Hydrogen	63.1	63.1/2.016 = 31.303	H_2O
Nitrogen	5.5	5.5/28.0134 = 0.196	N_2
Sulfur	0.1	0.1/32.060 = 0.003	SO_2
Oxygen	397.4	397.4/31.999 = 12.419	CO_2, SO_2, H_2O
Water	1325.6	1325.6/18.015 = 73.581	H_2O
Ash	12.0		
Sum	2325.6		

If we assume that 0.06 gS/kgdf escapes as SO_2 in flue gas, then the sulfur balance is:

MolS/kgdf	Mass, gS/kgdf	End Product
0.003	0.10	Available sulfur
−0.002	−0.06	SO_2
0.001	0.04	As ash

If we assume that unburned carbon in ash is 0.6 g/kgdf, then the carbon balance is:

	MolC/kgdf	Mass, gC/kgdf	End Product
	43.452	521.9	Available carbon
	−0.050	−0.6	Unburned
Sum	43.402	521.3	CO_2

If we assume stoichiometric combustion, then the oxygen balance is:

	$MolO_2$/kgdf	Mass, gO/kgdf	End Product
	12.419	397.4	Available oxygen
	−43.402	−1388.8	CO_2
	−0.002	−0.1	SO_2
	−15.651	−500.8	H_2O
Sum	−46.636	−1492.3	Oxygen requirement

If we burn the biomass with an air ratio of 1.25 (or 25% excess air), then the humid air demand is 1.25*1.4923/0.22925 = 8.1368 kg/kgds. In this the 0.22925 is the ratio of oxygen mass to total air mass.

If we look at the stoichiometric air requirement versus the lower heating value (LHV) Fig. 2.5 we obtain for LHV 20.0 MJ/kgdf 0.33 kg/MJ. Air demand is then 1.25*0.33*20.0 or 8.25 kg/kgdf. So using graphical estimation gives one close value to more accurately calculated value.

In reality much of the combustion proceeds so that intermediate gaseous carbon monoxide (CO) and hydrogen are formed. In particular, there is often not enough oxygen to reach the char surface. In the absence of oxygen the char tends to be gasified with carbon dioxide (CO_2) or water vapor (H_2O). Both carbon dioxide and water vapor react with char to form carbon monoxide CO.

$$C_{char} + CO_2 \rightarrow 2CO \qquad (2.13)$$

$$C_{char} + H_2O \rightarrow CO + H_2 \qquad (2.14)$$

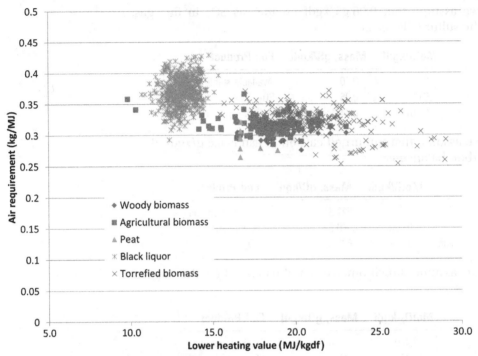

Figure 2.5 Stoichiometric air requirement kg air/MJ of biomass fuels.

The CO and H_2 are further oxidized to CO_2 and H_2O, respectively, further in the furnace where oxygen is available.

2.5.3 Mass Balance

Fluidized bed biomass boiler mass balance can be formed by looking at all flows to and from the boiler.

Solid input into the furnace through boiler enclosure is:

- fuel(s)
- sorbent (limestone)
- supplemental bed material (added sand)

Gaseous/liquid input into the furnace through the boiler enclosure is:

- air (primary, secondary, tertiary, leakage)
- auxiliary fuel (oil, natural gas)
- process streams (possible gas streams to be treated)
- steam (sootblowing, atomizing)

Solid output from the furnace through the boiler enclosure is:

- ash drain from the backpass
- ash drain from the external heat exchanger (loop seal)

- ash drain from the fluidized bed
- ash drain from the fabric filter (FF) or electrostatic precipitator (ESP)
- particulate solids leaving the stack

Gaseous/liquid output from the furnace through the boiler enclosure is:

- flue gas

Biomass solids, continually added to the furnace as fuel to be burned, contain ash. Therefore ash must be removed to keep the solids inventory stable. As note one should remember that input flows always add up to output flows. There is often an additional flow of limestone or dolomite to control sulfur emissions. The reaction product calcium sulfate is removed with bottom ash. As removed bottom ash from a fluidized boiler contains some of the bed inventory one must add fresh, sized sand to keep fluidized bed properties, especially the particle size, constant. Largest fraction of the total ash is fly ash that escapes from the furnace due to its small particle size. About 30–100% of ash is collected by the FF or ESP before the stack.

Continuing with the previous values we can calculate the ash and flue gas flows. If we assume there are 0.3 g/kgdf particulates escaping with flue gas and 3 g/kgdf sand addition, then for ash flow:

Mass (g/kgdf)	End Product
12.0	Ash in fuel
3.0	Sand addition
0.6	Unburned C in ash
−0.3	Dust in flue gas
0.0	Sulfur in bottom ash
15.3	Ash out

Similarly summing up all streams and assuming that 50 g/kgdf of sootblowing steam is used, then for the flue gas flow:

Mass (g/kgdf)	End Product
2325.6	Fuel
8136.8	Air
−0.3	Dust
−15.3	Ash
50.0	Sootblowing
10496.8	Flue gases

2.5.4 Heating Value

Fuel heating value is one of the most fundamental values in boiler calculations. Heating value is used as the basis for energy balances. Similarly in boiler dimensioning engineering, constants like the furnace heat load and heart heat release rate (HHRR) are specified using the fuel heating value. Flue gas production depends on the fuel heating value, Fig. 2.6. For the simple analysis of biofuels, combustion estimations based on the predetermined mass ratio of the flue gas to heating value can be done.

If not otherwise agreed, determination of the solid biomass heating value and analysis for the boiler design and performance test is done at an accredited laboratory. Typically oxygen content is not measured because of the measurement cost. Instead oxygen is assumed to make up the remainder to 100%. The fuel oxygen content can be analyzed to determine the total analytical error.

Higher and lower heating values in dry solid biomass fuel are related (Channivala, 2002). A higher heating value includes the

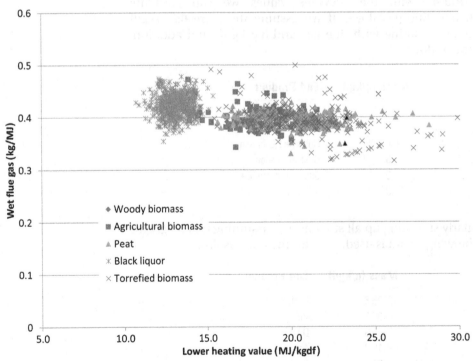

Figure 2.6 Stoichiometric flue gas from 1 kg of fuel at air ratio 1.2 divided by lower heating value, kg flue gas/MJ of biomass fuels.

condensation heat of water vapor to liquid water formed when hydrogen in the fuel is combusted to water vapor.

$$LHV = HHV - 2.443^*(M_{H_2O}/M_{H_2})^*X_H \qquad (2.15)$$

where

X_H is the hydrogen mass fraction in dry fuel

M_i is the mole weight of i, kg/kmol

Similarly the wet lower heating value and wet higher heating value can be expressed with dry heating values as

$$LHV_{wet} = LHV_{dry} * (1 - X_{H_2O}) - 2.443 * X \qquad (2.16)$$

$$HHV_{wet} = HHV_{dry} * (1 - X_{H_2O}) \qquad (2.17)$$

where

X_{H_2O} is the moisture content as mass fraction in wet fuel

The biomass hydrogen to carbon ratio is fairly constant. Therefore the biomass heating value is a function of the carbon content.

2.5.5 Heat Balance

Energy balances around the balance boundary are formed by noting that energy flows out in equal energy flows. Energy balances are the starting point of biomass boiler dimensioning. Heat transfer surfaces cannot be sized if some of the mass and energy flows are unknown. Balance calculations are also important as accurate energy and mass flows are needed to evaluate boiler plant economics and running costs. For boiler design and performance testing one can calculate heat balances based on the European standard EN 12952-15:2003 "*Water-tube boilers and auxiliary installations – Part 15: Acceptance tests.*"

Basically all heat balances need a reference state for thermodynamic properties. The term "*standard state*" is meant here to refer to the zero-point of enthalpy $P_n = 0.101325$ MPa and $T_n = 0°C$. It is recommended that balances are calculated using 0°C as the reference temperature. The balance reference temperature can also be some other temperature.

The energy input has two types of component. One is roughly proportional to the fuel flow, and the other does not directly depend on the fuel flow. An example of the first is air preheating. An example of the second is energy input by a flue gas recirculation fan.

Continuing with the previous values, we can calculate the heat to steam and the generated steam flow. To do that we need some additional definitions. One can read the amount of heat in steam (called enthalpy) from steam tables or the h–s figure provided in the Appendix. For now we just assume the enthalpies as input values.

Reference Temperature	0	°C
Higher heating value (HHV)	21390.0	kJ/kgdf
Moisture content	57.0	%
Hydrogen	6.31	mass-%
Air temperature to fans	30.0	°C
Air temperature to furnace	160.0	°C
Infiltration air	5.00	mass-%
Air c_p	1.03	kJ/kg°C
Sootblowing steam	From outside	
Sootblowing steam enthalpy	3054.8	kJ/kg
Heat of vaporization at reference temperature	2500.9	kJ/kg
Flue gas temperature	155.0	°C
Flue gas c_p	1.23	kJ/kg
Steam enthalpy corresponding to flue gas	2792.0	kJ/kg
Radiation and conduction	0.283	%
Unburned and others	0.300	%
Margin	0.500	%
Feedwater enthalpy	490.3	kJ/kg
Steam enthalpy	3360.7	kJ/kg
Blowdown enthalpy	1423.3	kJ/kg
Blowdown flow	0.1000	kg/kgdf

We can then calculate the input flows:

Item		kJ/kgdf	%
Lower Heating Value ($l = 2453$)	20006.7		110.7
Reference temperature correction	0		0.0
Water in fuel	−3251.7		−18.0
Heat in wet fuel		16755.0	92.7
Heat in Air			
incoming	0.95*8.136*1.03*(30-0)	239.7	1.3
heating	0.95*8.136*1.03*(160-0)	1038.6	5.7
infiltration	0.05*8.136*1.03*(30-0)	12.6	0.1
Outside sootblowing	0.05*(3054.8-2500.9)	27.7	0.2
Sum in		18073.7	100.0

The heat available can then be calculated by subtracting the losses from total heat input:

		kJ/kgdf	%
Wet flue gas	10.497*1.23*(155-0)	2001.2	11.1
Radiation and conduction	0.00283*18074	51.1	0.3
Unburned and others	0.003*18074	54.2	0.3
Marginal	0.005*18074	90.4	0.5
Sum		2196.9	12.2
Net to steam		15876.7	87.8

The efficiencies can then be calculated as:

Efficiency (higher heating value)	15877/(21390 + 18074 − 16755)*100	69.9
Efficiency (lower heating value)	15877/18074*100	87.8

The steam production can be found if we first subtract from heat to steam the heat required for heat blowdown and then divide with the enthalpy difference of the steam and feedwater:

$$(15877 − (1423.3 − 490.3)*0.1)/(3360.7 − 490.3) = 5.499 \text{ kg/kgdf}$$

2.6   Emissions

One of the main environmental concerns is the production of harmful emissions to air, i.e., pollution (OECD, 1974):

Pollution means the introduction by man, directly or indirectly, of substances or energy into the environment, resulting in deleterious effects of such a nature as to endanger human health, harm living resources and ecosystems, and impair or interfere with amenities and other legitimate uses of the environment.

In air pollution the natural characteristics of the atmosphere are changed when a chemical compound or some particulate matter is introduced to the atmosphere. Air pollution has long been one of the major environmental concerns and causes illnesses, especially respiratory diseases. Energy production, industry, and more and more transport are the greatest sources of emissions to air. Major air pollutants are:

- sulfur emissions: SO_2, S_{tot}
- nitrogen emissions: NO, NO_2
- particulate emissions: dust

They are emitted in large quantities due to human actions and processes in nature.

Adverse air quality affects the expected lifetime of humans. Often we refer to premature death (i.e., death occurred sooner than expected) to describe the most harmful effects of air pollution. Typical adverse effects of air pollution are respiratory diseases, cardiovascular diseases, throat inflammation, chest pain, and congestion. Particulate emissions, especially those from trace elements, can cause severe health problems (e.g., if heavy metals are introduced into the bloodstream). Sulfur dioxide and oxides of nitrogen form acidic components. These can lower the pH value of soils. Acidic soil can become unsuitable for plants. Smog and haze from nitrogen oxides reduce the amount of sunlight received by plants.

Natural processes cause significant air pollution. Large volcanic eruptions have resulted in short-term climate changes. Forest fires can spread smoke and very problematic pollutants to large areas. Wind causes erosion and thus particulates, which have been known to travel thousands of kilometers before depositing.

Almost all industrial nations have enacted legislation to regulate air. Many international bodies like the International Monetary Fund (IMF) have made their own requirements to apply globally (International Finance Corporation, 2008). Typically the target is to limit air pollution so that there is very little change in the nearby ecosystems. Usually large-scale release of pollutants is regulated and the boiler operator has to have an environmental permit. The polluter (i.e., the boiler owner/operator) needs also to monitor their actions to satisfy authorities that they are not polluting. As an example, typical emissions form woody biofuel combustion can be seen in Table 2.8.

In the United States the Clean Air Act was passed in 1963 to legislate against atmospheric pollution, especially as a result of the terrible smog that year. More importantly, on January 1, 1970 the threshold standards for air pollutants to protect human health were established by the US Environmental Protection Agency (EPA). Allowable emissions from large-scale biomass burning are thus looked at from two angles. The first is the best available technology (BAT) standpoint. This means that the authorities want the boiler operator to choose the most environmentally beneficial technical option—one that is widely available and in use. This also means that the technology chosen must be environmentally competitive even if the emissions do not cause major harm. Second, the authorities look at the ground-level exposure. This means performing atmospheric dispersion modeling in order to predict possible adverse changes in air quality. If the environment is already badly polluted, adding an additional load must be carefully considered.

Table 2.8 Typical Emissions from the Combustion of Woody Biofuel, mg/m^3N

Boiler Type	Grate	Bubbling Fluidized Bed	Circulating Fluidized Bed	Recovery
CO_2	150,000−~200,000	150,000−~200,000	150,000−~200,000	210,000−~240,000
CO	100−600	100−250	50−100	50−200
H_2S	3−20	1−20	1−5	0−4
SO_2	250−400	30−150	5−50	0−5
NOx	100−400	250−450	150−250	150−250
N_2O	0−1	4−8	5−10	0−1
HF	1−5	1−5	1−5	0
HCl	5−30	5−30	5−30	0−2
Dust, ESP	10−20	10−20	10−20	10−50
Dust, FF	1−5	1−5	1−5	N.A.
Dioxins and furans	<0.0001	<0.0001	<0.0001	<0.00001

ESP, electrostatic precipitator; *FF*, fabric filter.

The current EU regulatory framework uses IPPC BAT BREF documents for controlling pollution. For example, large biomass boilers need to look at the European Commission Integrated Pollution Prevention and Control (IPPC) Reference Document on Best Available Techniques for Large Combustion Plants (European Commission, 2006). This legally nonbinding but in practice strongly directing document is accompanied by a binding document called BAT Conclusions. In addition, boilers need to adhere to the EU Industrial Emissions Directive (IED) (European Council Directive, 2010). The IED lists maximum limits for many pollutants.

2.6.1 NOx Emissions

Air is mostly made of nitrogen. All living organisms contain nitrogen. Nitrogen is essential for the growth of biomass. Therefore nitrogen fertilizers are very common. Because biomass is made of living organic material, the nitrogen content of biofuels tends to be high, Table 2.9. There is lower nitrogen in the tree trunk, which is no longer growing, than in bark or branches, which are still growing.

Table 2.9 Nitrogen Content in Biomass Fuels

Biofuel	Percentage of Dry Mass
Wood	0.1–0.5
Bark	~0.5
Straw	0.5–1
Sewage sludge	~1
Black liquor	0.1–0.2
Refuse-derived fuel (RDF)	~1
Leather waste	~12

Nitric oxide (NO) is the main nitrogen oxide emission from solid biomass combustion (Tsupari et al., 2007). NO reacts further to nitrogen dioxide (NO_2) in the atmosphere. Nitric oxide is a colorless, odorless gas. Nitrogen dioxide is a reddish-brown gas with a pungent odor. Nitrogen oxides contribute to acid rain formation. Nitrogen dioxide affects the human respiratory system, aggravating asthma and other lung diseases. Major manmade releases of nitrogen oxides are for the large part caused by traffic (60%) and to a lesser degree by fuel combustion for energy (30%). In addition to NO, nitrous oxide (N_2O) is released during combustion. N_2O is a significant source of greenhouse gas emissions from human activities. About 30% of global N_2O is released from fossil fuel-fired heat and power and land use. The largest source of nitrous oxide is natural processes such as forest fires, biological processes and changes in land use.

NOx formation in boilers burning biomass fuels occurs through two main routes. In thermal NOx formation a high temperature is responsible for NO formation. This can be controlled by modern firing methods and often is not a big problem. No matter how we fire, about a third of nitrogen in solid biofuels is converted (via ammonia, e.g.,) to NO close to the entry point of that fuel into the furnace. Therefore a high level of nitrogen in fuel means that we must operate the furnace so that the NO that is created is reduced as much as possible.

NOx emission reduction in the furnace can be done by optimization of the feeding of air and fuel, increasing the number of air levels and utilizing flue gas recycling. Typically, extensive CFD calculations are done to find the optimum placement of air ports. After the furnace one can reduce NOx by feeding the ammonia by selective non-catalytic reduction (SNCR), by selective catalyst reduction (SCR) or by oxidation/reduction processes. The fluidized bed is operated at a much lower combustion temperature and subsequently NOx compounds are primarily from nitrogen in the fuel with negligible amounts of thermal NOx.

Flue gas recirculation can be used to lower NOx. It helps if the aim is to burn a wide variety of fuels. Flue gas recirculation can be used to maintain adequate gas flow. It also allows both combustion and steam temperature to be effectively controlled in a wide variety of operating conditions. Vainio et al. (2012) reported that only less than 10% of the available nitrogen in fuel was converted to NO emissions in a bubbling fluidized bed (BFB) boiler burning bark, sludge, and solid recovered fuel (SRF).

2.6.2 SO_2 Emissions

Sulfur is often found in fuel. Sulfur is one of the most common elements in the earth's crust. During burning it forms gaseous compounds. These fall to the ground with rain. Sulfur emissions cause acidification of soil. In solid biofuels (wood, straw, etc.) sulfur is often seen as inorganic contaminants (pyrite, SO_4^{2-}, elementary sulfur) and as organic sulfur, where sulfur is chemically bound with organic compounds forming the biomass. In gaseous biofuels (biogas) the sulfur is in the form of gaseous impurities. In liquid biofuels (bio-oil) the sulfur is found as organic sulfur.

During combustion the sulfur reacts readily with oxygen and forms sulfur dioxide (SO_2). Some reacts further to form sulfur trioxide (SO_3).

Sometimes one finds sulfuric acid (H_2SO_4) emissions in the flue gas. A fraction of the sulfur dioxide reacts with molecular oxygen in the upper furnace and forms sulfur trioxide (SO_3). During sampling or at the surface, SO_3 reacts further with water vapor to form H_2SO_4. In biomass combustion the sulfuric acid content has typically been negligible, but it can increase to be few parts per million (ppm) if the SO_2 concentration is several hundred ppm.

Sulfur dioxide emission can be reduced during combustion by the addition of lime (CaO), limestone ($CaCO_3$), or dolomite ($CaMg(CO_3)_2$) (fluidized bed firing) or sodium (occurs naturally with the burning of black liquor). Sulfur dioxide emissions can be decreased after combustion by wet or dry scrubbing. When limestone reaches typical fluidized bed temperatures, carbonate will decompose thus generating pores and dramatically increasing the reactive surface area.

$$CaCO_3 \leftrightarrow CaO + CO_2 \tag{2.18}$$

$$CaMg(CO_3)_2 \leftrightarrow CaO + MgO + 2CO_2$$

Calcium oxide then reacts with sulfur dioxide and further with oxygen to form gypsum:

$$CaO + SO_2 + \frac{1}{2}O_2 \leftrightarrow CaSO_4 \tag{2.19}$$

$$CaO + MgO + SO_2 + \frac{1}{2}O_2 \leftrightarrow CaSO_4 + MgO$$

A great advantage of fluidized bed combustors is that sulfur capture by limestone is fairly cheap and in general is a simple process to operate. Reacted material mostly leaves the furnace with the bottom ash.

2.6.3 Carbon Monoxide Emissions

Carbon monoxide (CO) is produced when combustion reactions are not fully completed, either through lack of oxygen or due to low mixing. Carbon monoxide is a colorless and odorless gas. All combustion sources, including motor vehicles, power stations, waste incinerators, domestic gas boilers, and cookers, emit carbon monoxide. Carbon monoxide is not poisonous but has a temporary effect on the human respiratory system. Carbon monoxide attaches itself to red blood cells, preventing the uptake of oxygen.

Carbon monoxide correlates with the oxygen content in the flue gases. Low excess oxygen increases CO formation. The higher the air ratio and the better the mixing, the lower is the CO emission. Operating with a high furnace temperature and a long residence time decrease CO emission. Because carbon monoxide emission behaves in a similar way to many other hydrocarbon emissions, it is often used for regulatory purposes as a signal for the overall efficiency of combustion. Thus regulating CO is often associated with imposing restrictions on hydrocarbon emissions.

It is typical for the NOx emission to increase with decreasing CO. From the total emission standpoint it is advantageous to run at some hundreds of ppm of CO to lower other harmful emissions and keep reasonable flue gas heat losses.

2.6.4 Total Reduced Sulfur Compounds

Reduced sulfur species, also known as total reduced sulfur (TRS) compounds, are smelly gases: hydrogen sulfide (H_2S), methyl mercaptan (MM), dimethyl sulfide (DMS), and dimethyl disulfide (Vakkilainen, 2005). The main source of TRS compounds is incomplete combustion in processes that produce sulfur emissions. Typically TRS emission is problematic with air emissions from pulping operations (kraft pulping odor).

2.6.5 Volatile Organic Compound and Organic Emissions

During solid biomass combustion literally millions of reactions are occurring simultaneously. Thermodynamics means that, counter to main reactions, large organic compounds can form to some degree. Some of the most noteworthy are polycyclic aromatic hydrocarbons (PAH), tar or condensable organic compounds, and total hydrocarbons (THC) or total organic carbons (TOC). Volatile organic compound (VOC) emissions typically do not include methane emissions, in which case they are known as nonmethane volatile organic compounds (NMVOCs).

VOCs are harmful because they help to increase the ground-level ozone concentration. Some VOC compounds are carcinogens. Motor vehicle exhausts and the chemicals industry are responsible for the highest share of VOC releases. A naturally occurring source of VOC is forest fires.

Polycyclic aromatic compounds are hydrocarbon compounds composed of several aromatic rings. Some carbon atoms may be substituted by nitrogen or sulfur, for example, giving heterocyclic PAHs. Typical PAHs found in flue gases or in pyrolysis or gasification product gases are those composed of 2 to 7 aromatic rings. PAH compounds are often carcinogenic. It is noteworthy that PAH emissions from small domestic biomass burning appliances are several decades larger than those from larger, industrial and utility biomass burning.

Dioxins and furans are formed in all combustion processes. Mostly the research on dioxins has concentrated on emissions from hazardous and municipal waste incinerators. Dioxin emissions can occur if high chlorine fuel is used in solid biomass boilers such as treated, varnished or PVC-coated wood. These can all produce high polychlorinated dibenzo-*p*-dioxin (PCDD) and polychlorinated dibenzofuran (PCDF) emissions (Lavric et al., 2004). If controlled conditions are maintained, then very few dioxins and furans are formed in biomass combustion.

2.6.6 HCl, HF

Fluorine (F), chlorine (Cl), bromine (Br), and Iodine (I) are often referred to as halogens. They are found in flue gases from the combustion of fuels and waste-streams. Chlorine is the most important of them because it is readily found in soil and water. Fluorine is mainly found in minerals such as CaF_2 and Na_2AlF_6, whilst bromine and iodine are mainly found in sea water (Br) and seaweeds (I). Ocean water contains chlorides, fluorides, bromides, and iodine.

Chlorine can be found in ESP dust as solid NaCl and KCl but it also exits in flue gases as HCl and Cl_2. Fluorine can mostly be found in flue gases as HF.

2.6.7 Dust Emissions

Small particles suspended in flue gas are called dust or particulate matter. Small particles are one of the most problematic air quality problems, especially in urban areas. Particles that are smaller than a hair's diameter can directly enter human blood through the respiratory system. Particulate matter comes from combustion. In addition solid particles are formed by road traffic when tires and asphalt wear down, by wind action causing erosion of solid material to dust, and by sea salt release when sea water droplets dry. Particulate matter is usually reported as the total solid mass per unit of flue gas. Additionally, small size particular matter is reported as PM10 or PM2.5 (particles with diameter less than or equal to 10 or 2.5 μm, respectively).

Particle formation in the furnace is a complex process, Fig. 2.7. First, inorganic compounds can vaporize from burning, hot particles and later condense to form small particles. Inorganic or organic reactions produce compounds (sulfates, carbonates), which later react. Typically there are plenty of nucleus particles formed through the fragmentation of biofuel ash during combustion. These particles grow through condensation or chemical reactions. They occasionally hit each other and coalesce into larger particles. This process is similar to raindrop formation. The higher furnace temperature in grate-fired boilers has been reported to promote higher fine particle and trace element concentration compared to that found in fluidized beds (Lind, et al., 2007).

Especially metals in particulate matter can cause respiratory illnesses. Heavy metals include antimony (Sb), copper (Cu), niobium (Nb), arsenic (As), gold (Au), selenium (Se), barium (Ba), iron (Fe), silver (Ag), beryllium (Be), lead (Pb), tellurium (Te),

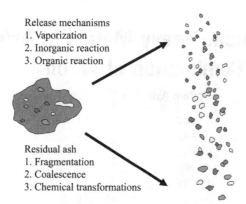

Release mechanisms
1. Vaporization
2. Inorganic reaction
3. Organic reaction

Residual ash
1. Fragmentation
2. Coalescence
3. Chemical transformations

Figure 2.7 Formation of particles during combustion.

cadmium (Cd), manganese (Mn), thallium (Th), chromium (Cr), mercury (Hg), zinc (Zn), cobalt (Co), nickel (Ni), vanadium (V), and other even more minor elements. Heavy metal emissions are known to end up in blood through the respiratory system. Heavy metal levels in biomass boiler dusts are typically low. The EPA in the United States lists 11 metals as hazardous air pollutants. These are Sb, As, Be, Cd, Cr, Co, Pb, Mn, Hg, Ni, and Se. Directive 2000/76/EC lists 12 heavy metals that require limits: Sb, As, Cd, Cr, Co, Cu, Pb, Mn, Hg, Ni, Th, and V. These lists differ but both contain the most problematic ones; cadmium and mercury. Manganese is included in both even though it is not a metal and is not very harmful.

Latva-Somppi (1998) found As, Cd, Pb, and Rb enriched in fly ash from sludge and wood co-firing. Na, K, S, and Cl formed condensed species on heating surfaces. The heavy metals can be divided into three groups based on their volatility. As, Cd, and Pb are highly volatile and vaporize completely or nearly completely. Se, and Cu are of medium volatility. Zn, Ni, Cr, and Co are of low volatility and occur only as a carryover, Table 2.10.

Kouvo and Backman (2003) found that even though Pb, Cu, Zn, and Mn occur in fuel, they mostly combine with other elements and stick to the surface of the sand in the bed. On the other hand, trace metal components captured in the bed are not necessarily thermally and/or chemically stable but may be released in specific conditions.

Dust can be controlled by an ESP, a filter, or a scrubber. When dealing with solid biomass, it is important to minimize dust emissions from the entire plant by the application of fugitive dust control to all material-handling equipment.

Table 2.10 Typical Heavy Metals in Ash From the Combustion of Wood

		Bottom Ash	Fly Ash
As	mg/kg	0.2–0.3	1–60
Cd	mg/kg	0.4–0.7	6–40
Co	mg/kg	0–7	3–200
Cr	mg/kg	40–60	40–250
Cu	mg/kg	15–300	100–600
Hg	mg/kg	0–0.4	0–1
Ni	mg/kg	40–250	20–100
Pb	mg/kg	15–60	40–1000
Se	mg/kg	0–8	5–15
V	mg/kg	10–120	20–30
Zn	mg/kg	15–1000	40–700

References

Alakangas, E., 2000. Suomessa käytettävien polttoaineiden ominaisuuksia (Properties of Finnish fuels). VTT Research Notes, 2045 VTT Energia, Espoo. 189 p. ISBN 9513856992 (in Finnish).

Aradhey, A., 2012. India—biofuels annual. USDA Foreign Agricultural Service, Global Agricultural Information Network, GAIN Report Number IN2081, 16 p.

Arbon, I.M., 2002. Worldwide use of biomass in power generation and combined heat and power schemes. Proceedings of the Institution of Mechanical Engineers, Part A: J. Power Energy, 41–57.

Barros, S., 2011. Brazil—biofuels annual. USDA Foreign Agricultural Service. Global Agricultural Information Network, GAIN Report Number BR110013, 32 p.

Beurskens, L.W.M., Hekkenberg, M., Vethman, P., 2011. Renewable energy projections as published in the National Renewable Energy Action Plans of the European Member States. Report ECN-E–10-069, Energy research Centre of the Netherlands, European Environment Agency, 234 p. ISBN 0919578152.

Caillat, S., Vakkilainen, E., 2013. Large-scale biomass combustion plants: an overview. In: Rosendahl, L. (Ed.), Biomass Combustion Science, Technology and Engineering. Woodhead Publishing Series in Energy, London, pp. 274–296. ISBN 9780857091314 (Chapter 8).

Channiwala, S.A., Parikh, P.P., 2002. An unified correlation for estimating HHV of solid, liquid and gaseous fuels. Fuel. 81 (8), 1051–1063.

Chase, M.W. Jr., 1998. NIST-JANAF thermochemical tables, Fourth ed. J. Phys. Chem. Ref. Data, Monograph 9, 1–1951.

Chum, H., et al., 2011. Bioenergy. Intergovernmental Panel on Climate Change, (IPCC), Special report on renewable Energy Sources and Climate Change Mitigation. Cambridge University Press, Cambridge, UK and New York, NY, 180 p (Chapter 2).

European Commission, 2006. Integrated Pollution Prevention and Control (IPPC), reference document on best available techniques for large combustion plants. European Integrated Pollution Prevention and Control Bureau, July 2006, 618 p.

European Council Directive, 2010. 2010/75/EU on industrial emissions (integrated pollution prevention and control), Europe, Brussels, of November 24, 2010, 156 p.

Frederick, W.J., Hupa, M., 1993. Combustion properties of kraft black liquors. Åbo Akademi, Department of Chemical Engineering, Report 93-3. 112 p.

Hess, R.J., Jacobson, J.J., Cafferty, K., Vandersloot, T., Nelson, R., Wolf, C., 2010. Country report—United States. IEA Bioenergy Task 40, 74 p.

Hyppänen, T., Raiko, R., 2002. Leijupoltto (Fluidized bed combustion). In: Raiko, R., Kurki-Suonio, I., Saastamoinen, J., Hupa, M. (Eds.), Poltto ja palaminen. International Flame Research Foundation, Suomen kansallinen osasto, Jyväskylä, pp. 490–521. (in Finnish) ISBN 9516666043.

International Energy Agency, 2014. World Energy Outlook 2014. International Energy Agency, Paris, France, 748 p. ISBN 9789264208056.

International Energy Agency, IEA energy statistics, 2012. Available from: <http://www.iea.org/stats/index.asp> (Accessed 22.05.12).

International Finance Corporation, 2008. Thermal power: guidelines for new plants. Part of World Bank Group Environmental, Health, and Safety Guidelines, December 19 2008, 33 p.

Järvinen, M., Zevenhoven, R., Vakkilainen, E., 2002. Implementation of a detailed black liquor combustion model for furnace calculations. IFRF Electronic Combust J, Article Number 200206, June 2002, 34 p.

Jenkins, B.M., Baxter, L.L., Miles Jr., T.R., Miles, T.R., 1998. Combustion properties of biomass. Fuel Process Technol. 54, 17–46.

Khan, A.A., 2007. Combustion and co-combustion of biomass in a bubbling fluidized bed boiler. Ph.D. Thesis, Technische Universiteit Delft, Netherland, 224 p. ISBN 9789090219646.

Kouvo, P., Backman, R., 2003. Estimation of trace element release and accumulation in the sand bed during bubbling fluidised bed co-combustion of biomass, peat, and refuse-derived fuels. Fuel. 82 (7), 741–753.

Kuparinen, K., Vakkilainen, E., Heinimö, J., 2014. World's largest biofuel and pellet plants—geographic distribution, capacity share, and feedstock supply. Biofuels Bioprod Bior. 8 (6), 747–754.

Lamers, P., Junginger, M., Hamelinck, C., Faaij, A., 2012. Developments in international solid biofuel trade–an analysis of volumes, policies, and marketfactors. Renewable and Sustainable Energy Reviews. 16 (5), 3176–3199.

Latva-Somppi, J., 1998. Experimental studies on pulp and paper mill sludge ash behavior on fluidized bed combustors. Ph.D. Thesis, VTT Publications 336, Technical research centre, VTT, Espoo, 155 p. ISBN 9513852148.

Lavric, E.D., Konnov, A.A., De Ruyck, J., 2004. Dioxin levels in wood combustion—a review. Biomass Bioenerg. 26 (2), 115–145.

Lind, T., Hokkinen, J., Jokiniemi, J.K., 2007. Fine particle and trace element emissions from waste combustion—comparison of fluidized bed and grate firing. Fuel Process Technol. 88 (7), 737–746.

Mantau, U., et al., 2010. EU wood—Real potential for changes in growth and use of EU forests. Final report. Hamburg, Germany, June 30, 2010, 160 p.

Miles, T.R., Miles Jr., T.R., Baxter, L.L., Bryers, R.W., Jenkins, B.M., Oden, L.L., 1995. Alkali deposits found in biomass power plants - A preliminary investigation of their extent and nature. Summary report for National Renewable Energy Laboratory, NREL Subcontract TZ-2-11226-1, 122 p.

MME, 2011. Brazilian Energy Balance, 2011. Ministry of Energy and Mines, Brazilian Federal Government.

OECD, 1974. Recommendation of the Council on Principles concerning Transfrontier Pollution. Organisation for Economic Co-Operation and Development, November 14, 1974—C(74)224.

Parikka, M., 2004. Global biomass fuel resources. Biomass Bioenerg. 27 (6), 613–620.

Ranz, W.E., Marshall, W.R., 1952. Evaporation from drops. Chem. Eng. Prog. 48, 141–146.

RES Electricity Directive, 2001. Directive 2001/77/EC of the European Parliament and of the Council of 27 September 2001 on the promotion of electricity produced from renewable energy sources in the internal electricity market. Official Journal of the European Communities, L. 44 (283), 33–40, 27.10.2001.

Saastamoinen, J., 2007. Simplified model for calculation of devolatilization in fluidized beds. Fuel. 85 (17-18), 2388–2395, December 2006.

Strömberg, B., 2006. Fuel handbook. Technical Report, SVF/971 Stiftelsen för Värmeteknisk Forskning, Stockholm, Sweden, 105 p.

Teir, S., 2004. Steam boiler technology, Energy Engineering and Environmental Protection publications. 2nd ed Helsinki University of Technology, Department of Mechanical Engineering 215 p. ISBN 9512267594.

Thrän, D.; Fritsche, U.; Hennig, C.; Rensberg, N. and Krautz, A., 2012. Country report—Germany. IEA Bioenergy Task 40, 55 p. Available at: <http://bioenergytrade.org/reports/country-reports-2010/index.html>.

Tsupari, E., Monni, S., Tormonen, K., Pellikka, T., Syria, S., 2007. Estimation of annual CH_4 and N_2O emissions from fluidised bed combustion: An advanced measurement-based method and its application to Finland. Int. J. Greenhouse Gas Control. 1 (3), 289–297.

Vainio, E., Brink, A., Hupa, M., Vesala, H., Kajolinna, T., 2012. Fate of fuel nitrogen in the furnace of an industrial bubbling fluidized bed boiler during combustion of biomass fuel mixtures. Energy Fuels. 26 (1), 94–101.

Vakkilainen, E.K., 2005. Kraft Recovery Boilers—Principles and Practice. Suomen Soodakattilayhdistys r.y., Valopaino Oy, Helsinki, Finland, 246 p. ISBN 9529186037.

Vakkilainen, E., Kuparinen, K., Heinimö, J., 2013. Large industrial users of energy biomass. IEA Bioenergy Task 40, September 2013, 75 p.

Walter, A., and Dolzan, P., 2012. Country report—Brazil. 2012 IEA Bioenergy Task 40, 65 p. Available at: <http://bioenergytrade.org/reports/country-reports-2010/index.html>.

Williams, G.H., 1992. Fuel from biomass. Chem. Eng. News. 70, 33–47.

WWF, 2011. The Energy Report: 100% Renewable Energy by 2050. WWF International, Ecofys and OMA, Gland, Switzerland, OMA, 256 p. ISBN 9782940443260.

Zevenhoven-Onderwater, M., Blomquist, J.-P., Skrifvars, B.-J., Backman, R., Hupa, M., 2000. The prediction of behaviour of ashes from five different solid fuels in fluidised bed combustion. Fuel. 79 (11), 1353–1361.

3

BOILER PROCESSES

CHAPTER OUTLINE

Steam Generation from Biomass. DOI: http://dx.doi.org/10.1016/B978-0-12-804389-9.00003-4

This chapter looks at how one can decide what kind of boiler is needed and how to calculate the boiler's main mass and energy balances. First we explore the boiler selection process, selection of main parameters, placement of heat transfer surfaces, calculation of heat loads, advanced cycles, main processes, and determination of boiler efficiency. The main tasks are the placement of heat transfer surfaces, determination of the flows of the main components, and taking the effects of pressure and temperature into account.

3.1   Boiler Selection Process

The designer/owner/operator needs to consider several factors when determining the type and configuration of the boiler. It is often difficult and costly to change them during the project so these values need to be considered carefully. The main tasks are:
1. Determine the steam/power requirement.
2. Determine the fuels available and their usability.
3. Determine possible locations to place the boiler.
4. Compare different types of equipment.
5. Anticipate future needs.
6. Permits.

The selection process is influenced by many additional factors (Teir, 2004). The main ones are applicable emission requirements, the project's time schedule, the standards/pressure vessel code to be used, and reliability. Almost always, several iterations need to be done to arrive at the final selection. Improving thermal efficiency requires extra investments. These investments can be partially or totally offset by savings in operating costs. Annual costs of owning and operating a plant are a sum of the annual charges for capital, fuel, maintenance, manpower, ash, and waste disposal (Advances, 1986). The best configuration thus depends on actual site conditions.

3.1.1   Steam/Power Requirement

The first task is to determine the required power output from the generator and then from the boiler. This necessitates initial rough estimates based on typical efficiencies. As the boiler design process is an iterative process, the actual fuel demand (boiler size) is known only after the boiler design has been finalized. However, investment comparisons can be made based on typical designs.

Based on the required electric power output and the possible requirement of process/district heating, the heat input to the boiler can be calculated. With the required heat input the fuel input can be calculated based on boiler efficiency. Then the thermal output of the boiler as well as the fuel input flow is known.

3.1.2 Fuel Availability

Fuel availability is one of the deciding figures in the steam plant selection process. The price of biofuel typically depends heavily on transport costs. Therefore local or nearby fuels are often favored. In particular, biofuels such as bark and chips have high transport costs. Their ash content is low, but moisture content is often 50% or more. Typically for wood-based fuels the economic transport distance is no more than 50–150 km.

When examining the availability of any specific fuel, one must look at the whole chain. Each fuel requires different types of harvesting and transport equipment, handling, feeding, and combustion systems, and specific flue gas and ash treatment systems (Raiko et al., 2003). Emission requirements dictate the use of specific fuels or fuel blends (e.g., low sulfur fuel) in specific types of steam generators.

Changing the fuel specifications typically changes the operation and generated emission levels. This might require new permits and in some cases costly modifications to the boiler.

3.1.3 Locations

If the steam generator is for utility use, the location can be chosen from a wide area. Preferably the location would be near the fuel supply and near the users of energy. Often one or the other has to be chosen.

The biofuel boiler location has to be chosen so that the fuel transport, possible dusting, noise from fuel loading and handling, visual impact, etc. are considered. Then a location that minimizes the adverse effects without increasing the plant costs excessively is chosen.

3.1.4 Boiler-Type Classes

Boilers, like all types of equipment, have fundamentally different-looking constructions depending on the boiler size, use, and fuel type. The main boiler usage classes are utility, industrial, heat recovery steam generators (HRSG), and process. Boilers are also classed based on the chosen firing process: grate, fluidized bed, and recovery. In particular the chosen operating pressure will have a large effect on the type of boiler. The available heat source (i.e., fuel type) and firing method dictate the fuel handling equipment and affect the types and placing of the heat transfer surfaces, in particular how the boiler furnace looks.

When determining the boiler type to be purchased, one must look at the cost of each type of power station. Based on available funding and environmental performance the best alternatives can be determined. It is often difficult to continue without actually getting preliminary quotes for delivery times and conditions based on the chosen boiler location. Finally the required load characteristics are still the main determining factor. Each combination of fuel and boiler type will affect the maintenance cost. Each fuel has unique corrosion and erosion properties.

3.1.5 Boiler Purchasing

The main purchase is the pressure vessel with the steel structure. Boiler purchasing requires also choosing and buying the boiler's auxiliary equipment, such as fuel handling equipment and storage, pumps, fans, and emission reduction equipment. Typically a building contractor does the foundations. Often another contractor makes the steel structure, and a dedicated contractor builds the stack.

Because of the large sums involved, there has to be tight financial control over the whole project. Loans must be scheduled, and payments should be given only against work done and purchases made. The cost of the project's capital can be significant. In addition the permit process should be closely monitored. A boiler should be purchased with environmental performance that will be acceptable also in the future. Timing of the permit process is also crucial.

Boiler equipment should be selected based on satisfactory expected life span and acceptable maintenance costs. This decision can be based on satisfactory previous references, for example. All equipment should have adequate durability. The equipment must be placed so that it is accessible for inspection and repair. When making the purchasing decision, the availability of spare parts should be investigated. In summary, safe, reliable operation is the main target, with pleasing working conditions.

3.1.6 Permits

Steam generation requires permits to operate. Zoning requirements place the boiler in an area that is reserved for energy generation and does not cause harm to areas where people live. Boiler equipment needs to be authorized for use. The boiler must be made according to the required pressure vessel code. The electrical system and instrumentation need to adhere to applicable laws.

Operating permits are often also needed. Environmental permits are used to regulate loads of emissions to the environment. These include gaseous effluents, effluents to the water (heat), and dumping of solid waste (ash). Boiler operators are often codified. Chemical laws need to be followed in handling streams.

3.2 Choice of Main Design Parameters

The need for electrical energy follows the increase in development of gross national product. With higher unit capacities the specific investment can be made smaller. The development of materials and manufacturing methods has made possible the use of higher and higher main steam pressures and temperatures. This has increased the steam generation efficiencies, saving fuel and the environment.

The main design parameters are the steam generation capacity, main steam temperature, and main steam pressure. Increased pressure increases efficiency but requires thicker tubes. Increased temperature increases efficiency, but increases corrosion risk and requires more expensive materials. The choice of fuel and firing type leads to a specific boiler design. Therefore the intended use (utility, industrial, HRSG), the process in addition to the heat source, and the fuel and firing method will lead to some typical furnace types and positions as well as the placement of the heat transfer surfaces.

3.2.1 Increasing Main Steam Pressure and Temperature

Main steam temperature increased until the 1970s. This was possible because of improvements in tube materials and better understanding of steam generation. Now we are in an era of no real increase. Good, economical materials above 550°C are not yet available. Early trials (in the 1930s) of once-through boilers used pressures above the triple point of water. Pressure then stabilized to correspond to 540–550°C main steam temperature where corrosion of the highest superheaters was acceptable with reasonably priced materials.

The quest for yet higher steam pressures and temperatures continues. Examples are the European AD 700 research program (Kjær et al., 2002). It aims for a main steam temperature of 700°C, pressure 37.5 MPa, and electricity producing efficiency of over 50%. The main problem still remains the same. Where can one get high temperature, weldable, tough, and corrosion-resistant materials? Similar R&D programs are funded in the United States by the Department of Energy (DOE) and in Japan by the Ministry of International Trade and Industry (MITI).

The economic driving force behind these schemes is somewhat lacking. Adopting these values in the newest boilers gives a gain in electricity production of only a few percent. It seems doubtful that these high values will be widely adopted very fast. It should be noted that the main increase to electricity-generating efficiency comes from increasing temperature not pressure. By using reheating or double reheating, one can push conditions after final expansion further in the steam turbine, increasing electricity generation.

Typically one fixes the main steam values at the beginning of the project. Changing main steam pressure or temperature during the project is costly. Changing them during the life time of the boiler is often uneconomical and in most cases not even practically possible.

3.2.2   Placement of Heat Transfer Surfaces

A boiler generally has a furnace for evaporating the water, superheating to increase the final steam temperature, economizers to improve efficiency and air preheaters. Heat transfer surfaces are generally placed in counter-current fashion. The notable exception is the furnace, which is typically the first heat transfer surface in the gas flow direction. Often also some superheaters are placed in parallel fashion to decrease corrosion. As a general rule, placing radiative heat transfer surfaces, which have high temperature differences, in parallel fashion is not economically problematic.

During the steam generator design the engineer positions the surfaces and then calculates the heat loads to each surface. The surface types are chosen, based on experience. Then the surfaces are dimensioned. The manufacturing cost is calculated, based on these dimensions. If the cost is reasonable and manufacturing is possible, then the placement of the heat transfer surfaces was successful. If not, then a new variation must be tried.

3.2.3   Evaporation

In the first steam generators, phase change consumed most of the heat. Phase change is called evaporation or boiling. The whole process of evaporation from saturated water to saturated steam takes place at an almost constant temperature. At low heat flux conditions this temperature is called the saturation temperature and can be found, e.g., from steam tables (Schmidt, 1989). Typically most of the evaporation is done in the furnace before the nose. Some evaporation is done on the upper walls of the furnace. The heat transfer coefficient for boiling in natural circulation boilers is in the range of 30,000–120,000 W/m^{2}K (Fig. 3.1).

3.2.4   Superheating

An increase of steam temperature is called superheating. To control the final steam temperature water is sprayed onto

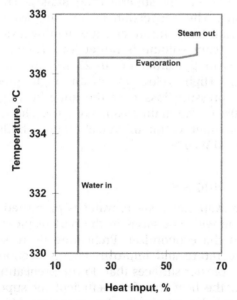

Figure 3.1 Conversion of water to steam. (Note the figure *y*-axis as temperature remains constant while water converts to steam.)

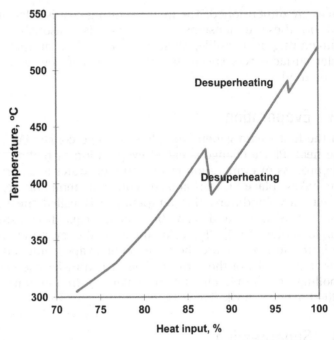

Figure 3.2 Superheating.

the steam between the superheating stages. This is called desuperheating. The conversion of spraying water to steam lowers the steam temperature causing the downward slopes in Fig. 3.2. The steam volume is rather low. Therefore we need high velocities in the order of 10–25 m/s to create efficient heat transfer. High velocities mean high pressure loss. Typically the pressure loss over the superheating section is more than 10% of the main steam pressure. The heat transfer coefficient for superheating in typical biomass boilers is in the range of 50–500 W/m^{2}K.

3.2.5 Economizers

Before the evaporative stage, water is preheated. Preheating is usually done with flue gases in the part of the heat transfer surface called the economizer. Preheating decreases the final flue gas temperature and improves energy economy. (This is why the heat transfer surfaces that do the preheating are called economizers.) the heat transfer coefficient for superheating in typical biomass boilers is in the range of 50–150 W/m^{2}K (Fig. 3.3).

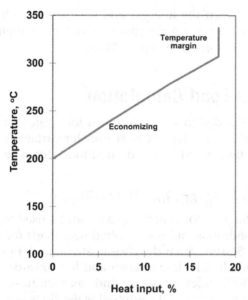

Figure 3.3 Preheating or economizer.

Ideally preheating is done up to saturation temperature. In practice a margin for evaporative temperature is needed to prevent boiling. If boiling occurs, unstable water surges can seriously harm boiler operation and even damage it. In practice, a temperature margin of about 30°C at maximum continuous rating (MCR) is often used.

3.2.6   Air Preheaters

Air preheating transfers heat from cold flue gases to heat the incoming air. This means a higher amount of heat per unit of produced flue gas in the furnace and thereby higher combustion temperatures. Air preheating thus has a dual purpose: higher temperatures in the furnace to improve reactions and improved steam generating efficiency.

There are two main types of air preheaters: gas to air and steam or water to air. If flue gas transfers heat to air, then almost invariably the air heat transfer surface is placed as the last heat transfer surface. The main problem with flue gas to air heat transfer is that as both media are gas, it is hard to reach overall heat transfer coefficients much over 20 W/m²K.

Air preheating also means smaller heat transfer surfaces. Increased flue gas temperatures mean a larger temperature difference between the flue gas and steam/water. On the other hand, air preheaters require more capital investment. In large units air preheating can be up to 300°C

3.3 Heat Load Calculation

Steam boiler design starts with heat load calculation. In heat load calculation each type of heat transfer surface is assigned a load or heat flow based on the desired total steam generation.

3.3.1 H-p Diagram for Water/Steam

An H-p diagram for water, Fig. 3.4, can be used to determine operating conditions and the required heat loads for heat transfer surfaces. Starting from the chosen steam outlet pressure and temperature, a line is drawn down and left towards the desired feedwater (FW) inlet pressure and temperature. Enthalpies where saturated steam and saturated water lines are crossed are marked.

The first estimation of the heat required for transforming water to steam is the difference between the points where the line crosses saturated steam and saturated water lines. As can be seen, there is no phase change above critical pressure

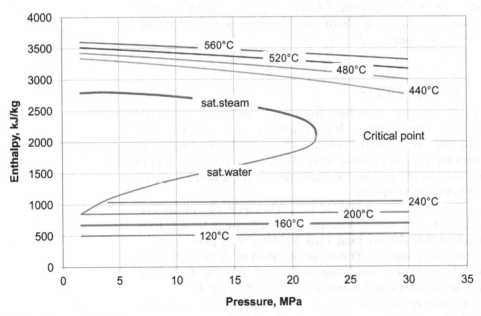

Figure 3.4 H-p diagram for water.

(22.12 MPa). This means no evaporative surfaces are needed. Water behaves continuously in this region. In general, a lower operating pressure means a larger latent heat requirement for the phase change. A higher operating pressure means more superheating and preheating are required.

The first estimation for the heat required to superheat steam is the enthalpy difference between the main steam outlet point and the point where the line crosses the saturated steam line, Fig. 3.9.

With higher pressures the portion of heat required for super-heating increases, up to critical pressure, Fig. 3.5. With supercritical cycles the temperature increase from enthalpy corresponding to critical temperature and pressure decreases with increasing pressure. As can be seen, the enthalpy needed around critical pressure forms an unstable saddle point. Controlling the final temperature is a demanding task there. This is one of the reasons why few boilers operate at close to critical pressure.

The first estimation for the heat required to bring feedwater to boiling is the enthalpy difference between the points where the line crosses the saturated water and the feedwater inlet point. With higher pressures the portion of heat required for economizers becomes larger.

The temperature—heat input profile can then be formed based on the heat of each part of the steam—water cycle. Fig. 3.6 shows a real temperature—heat input profile for a

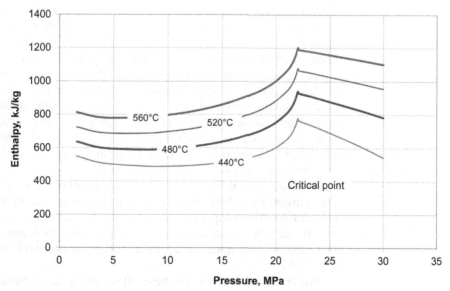

Figure 3.5 Enthalpy for ideal superheating as a function of main steam values.

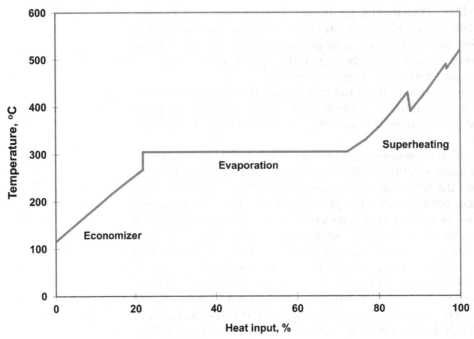

Figure 3.6 Temperature—heat input profile for a natural circulation boiler.

natural circulation boiler. The heat required for evaporation is smaller than what would be given by an initial estimate. Economizing is not usually completed. Water below some 30°C is brought to be evaporated. This is to avoid flow instabilities caused by pressure swings when steam bubbles are formed. Economizer heat is then replaced by evaporative heat.

There are sharp edges in the superheating part in Fig. 3.6. Usually water is sprayed between the superheating sections to control steam temperature. During spraying the steam temperature drops as sprayed water is evaporated. This means that from 30 to 50°C superheating must be redone. This increases the heat required for superheating and decreases the heat required for economizing and evaporating.

In a supercritical cycle no evaporation occurs. Fig. 3.6 shows the temperature—heat input profile of a real supercritical boiler superimposed on top of a T-h diagram. As can be seen, there is no constant temperature part in the process. Therefore, during fast process changes, local tube temperatures will fluctuate (Fig. 3.7).

Supercritical boilers are favored as large units because for each MWth they require less steel. Supercritical boilers do not

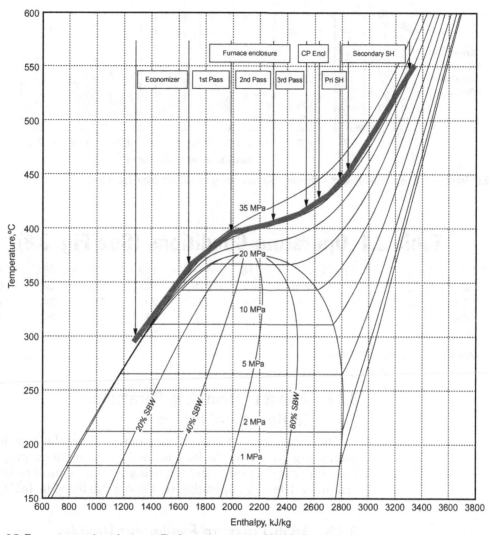

Figure 3.7 Temperature—heat input profile for a pure supercritical boiler. Redrawn from Smith, J.W., 1998. Babcock & Wilcox Company supercritical (once through) boiler technology. Babcock & Wilcox Technical paper, BR-1658, Barberton, OH, 9 p. (Smith, 1998).

have a drum and use smaller diameter tubes in the furnace than natural circulation boilers. Supercritical boilers, however, often have higher own power use because of higher operating pressure and pressure losses during evaporation. They also require more careful operation during start-up, shut-down, and partial load.

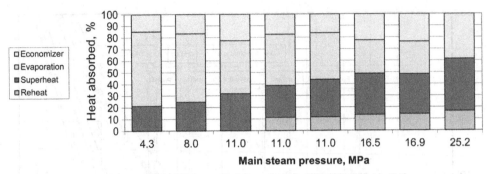

Figure 3.8 Fractions of heat needed with different main operating conditions (see Table 3.1).

Table 3.1 Operating Conditions (See Fig. 3.8)

Pressure, MPa	4.3	8.0	11.0	11.0	11.0	16.5	16.9	25.2
Temperature, °C	460	480	570	525	540	570	570	600
FW temperature, °C	174	196	204	230	230	254	246	273
RH pressure, MPa	—	—	—	2.8	2.8	4.1	3.2	4.9
RH temperature, °C	—	—	—	513	541	541	568	568

Abbreviation: FW, feedwater; RH, reheat.

3.3.2 Effect of Operating Parameters to Heat Input Profile

The steam generator main parameters affect the heat input profile. Fig. 3.8 shows the heat needed for economizer, evaporation, superheating (SH), and reheating (RH) on typical commercial units. The main operating conditions are listed in Table 3.1.

3.3.3 High-Pressure Feedwater Heaters

Electricity generation can be increased by modifying the basic Rankine cycle. Advanced cycles use extra heat transfer surfaces and more complicated processes to increase unit electricity-generating efficiency. Often-used features to increase electricity generation are high-pressure feedwater heaters and reheaters. Because of the added complexity and cost, it pays to use these features only in large unit sizes.

In advanced cycles, steam generation can be increased by increasing feedwater heating. This means steam heat exchangers that increase the feedwater inlet temperature. The number and extent of feedwater heating is solved by iterative means and depends mainly on unit size and manufacturing costs.

When steam is extracted from the turbine to increase water preheating, that steam has expanded in the turbine and generated electricity. The flow of steam per unit heat input increases as less heat is needed to heat feedwater to steam. We note that the higher the main steam pressure, the higher the feedwater inlet temperature tends to be.

3.3.4 Reheating Cycle

If the chosen cycle includes reheating, a new line is drawn from the chosen reheater outlet temperature and pressure down and right towards the desired reheater inlet temperature and pressure. The inlet condition is mainly dictated by the expansion in the turbine. The first estimation for the heat required to reheat steam is the enthalpy difference between the reheater steam (RS) outlet and inlet points.

In a quest to increase electricity-generating efficiency over 40%, reheating cycles became popular. A reheating cycle is shown in Fig. 3.9. After the first expansion 1−7 a new superheat 8−9 is done. For increased efficiency the steam from the final expansion 9 can be used to preheat the feedwater 4. Looking at the figure, it is easy to estimate that the additional process 8−9−2−7 has a higher electricity-generating efficiency than

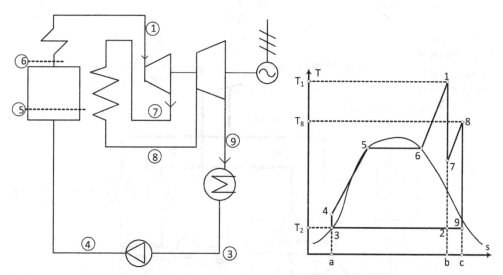

Figure 3.9 Ideal reheating cycle. (Left) Process flow diagram; (Right) same process displayed in T-s diagram 3-4 pumping, 4-5 economizing, 5-6 evaporating, 6-1 superheating, 1-7 expansion in turbine, 7-8 reheating, 8-9 expansion in turbine, 9-3 condensation. After Bidard, R., Bonnin, J., 1979. Energetique et turbomachines. (Theory of energy and turbomachinery). Eyrolles, Paris, 742 p. ISSN 0399-4198 (in French) (Bidard and Bonnin, 1979).

the base process 2–3–4–5–6–7. Thus the efficiency of the whole cycle improves. With higher pressures the portion of heat required for reheating increases.

3.4 Cogeneration

Often in the process industry and in district heating power plants a large portion of the steam from the turbine is used for process heating purposes. These kinds of power plants are called cogeneration power plants and the power they produce is called cogeneration power. In district heating the back pressure steam heats water circulating in the district heating network. Another example of cogeneration is the bark boiler in pulp mills. The wood wastes in the pulp and paper industry today commonly find a use in cogeneration power plants. The chemical energy in the wood waste is released by burning in a steam boiler (Huhtinen et al., 1999).

Fig. 3.10 shows the flow diagram of a typical cogeneration plant. Fuel and air burn in a steam generator and produce steam.

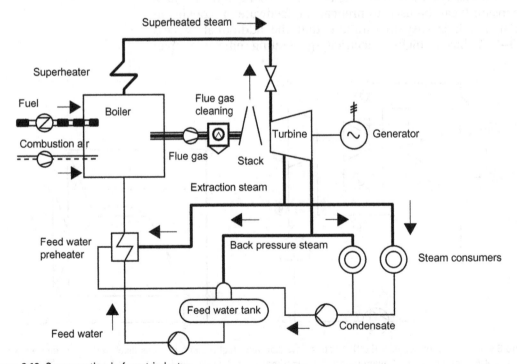

Figure 3.10 Cogeneration in forest industry.

The steam goes to the turbine. Part of the steam is extracted from the turbine at pressure ranging from 1.0 to 1.5 MPa. This part is called medium pressure steam. Part of the steam is extracted from the turbine at pressure ranging from 0.3 to 0.6 MPa. This part is called back pressure steam. In some cases there can be up to four different pressure levels (Huhtinen et al., 1999). The pressure levels are chosen to correspond to the required heat users. Often some steam expands further from 0.005 to 0.012 MPa. This part is called the condensing power.

Often the industrial requirement of steam for heating allows the production of electricity though cogeneration since the marginal investment cost for electricity is low when heating is needed anyway. Industrial cogeneration plants often have low electrical efficiencies (~20%) but a very high overall efficiency (+80%).

3.5 Main Processes

The steam generation process is generally divided into smaller subprocesses based on the media used. Often this division in subprocesses is reflected on piping and instrumentation diagrams (P&ID). Typically the main processes are:
- Air system: Parts of boiler connected with air flow
- Flue gas system: Parts of boiler connected with flue gas flow
- Steam–water system: Components connected with steam and water flow
- Ash system: Components connected with ash flow
- Fuel system: Components connected with fuel flow

3.5.1 Air System

The combustion air is led to the furnace. The air system consists of components that are used to get air to the boiler furnace. Manufacturing reasons dictate that the furnace is kept close to the surrounding pressure. Therefore blowers are needed to drive air into the furnace. Ducting must be done from where air is drawn to the blowers. Silencers are often used to lower the noise of the air entry. Ducting continues between the blowers and the furnace. Air flow is typically controlled by dampers in addition to the blower operating point control. Air flow, pressure, and temperature are measured. To improve combustion the air is heated. Heat exchangers are common where exiting flue gas heats air.

3.5.2 Flue Gas System

When biomass is burned in the furnace, then flue gases are formed. The flue gas system consists of components that are used to get flue gas from the boiler furnace to exit through the stack. Similarly to the air system, flue gas fans draw air from the furnace to improve the draft of the stack. Flue gas ducts connect pieces of equipment. Electrostatic precipitators and other emission reduction equipment are employed to decrease pollution.

3.5.3 Steam—Water System

Heat receiving media in most biomass boilers consist of a steam—water mixture. The steam—water system consists of components that are used to convert water to steam. Feedwater is pumped from a feedwater tank to the boiler. Feedwater is heated in the economizers and enters the drum. Saturated water flows downward in tubes called downcomers to be divided and distributed. Evaporating water rises up the wall tubes. Steam is formed as heat absorbed by the tubes results in phase change. Steam is separated from water in the drum and led to the superheaters. Feedwater is injected between the superheaters to control the final steam temperature. A saturated water stream called blowdown is bled from the drum to purge impurities from the boiler water.

3.5.4 Ash System

When we burn biomass not everything in the incoming biomass is converted to flue gas. The part of the fuel that is not combusted forms ash. The ash system consists of components that are used to convey ash from various locations of the boiler to disposal. The ash system requires ash transport equipment such as screws and conveyors. There are hoppers where ash is collected. Rotary feeders are used to control ash flow. Ash can be handled dry or sluiced and pumped away.

3.5.5 Fuel System

Combustion requires fuel. The fuel system consists of components that are used to transport biomass fuel from storage to the furnace for combustion. The fuel system needs to have flow measurement. Typically there is more than one insertion point of the fuel to the furnace. Fuel system design is usually thoroughly covered by standards and recommendations. This is because if a fuel system malfunctions, a fire or explosion outside or inside the furnace could occur with loss of life and property.

3.6 Determination of Boiler Efficiency

Boiler efficiency is a measure of the goodness of the chosen process and equipment to transfer combustion heat to the heat in steam. Boiler efficiency can be defined as the ratio of the useful heat output to the total energy input.

$$\eta = \frac{Q_{abs}}{Q_{in}} \qquad (3.1)$$

where
  η is boiler efficiency
  Q_{abs} is the useful heat absorbed (heat transferred to steam)
  Q_{in} is the heat and energy input into the boiler

Typical boiler efficiencies range from about 90% for the best solid biomass fuel boilers to close to 95% for oil- and natural gas-fired boilers, Table 3.2. The main reason for the poorer performance of biofuels is the high moisture content of the fuel, which increases flue gas losses.

To determine the efficiency of a boiler, a system boundary must be defined and the energy flows that cross the boundary need to be resolved. System boundaries should be chosen so that it is possible to define all energy and mass flows in and out with sufficient accuracy. In practice, many minor flows are usually neglected. When determining the boiler efficiency, all internal reactions and recirculation can be neglected. Determination

Table 3.2 Typical Boiler Efficiencies Calculated According to EN 12952-15

Fuel	Efficiency, %
Natural gas	94–95
Oil	92–95
Coal	88–92
Wood chips	87–91
Bark	85–90
Peat	85–89

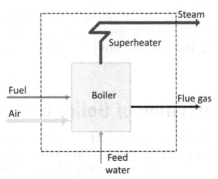

Figure 3.11 Simplified boiler schematic.

of boiler efficiency can be done with only flows though system boundaries.

Fig. 3.11 shows a simple boiler process. Fuel and the required amount of combustion air are fed into the boiler. The fuel reacts with the oxygen in the boiler and flue gas is drawn out. Released heat is captured by the water pumped into the boiler, transforming it into the steam that flows from the boiler.

It is clear that the system boundaries can be drawn in many ways. For example, the boiler house might form a convenient system boundary. It has been pointed out that losses and heat input caused by fans, blowers, and pumps should not affect the boiler efficiency. On the other hand, forced circulation pumps, flue gas recirculation fans, and other internal process devices should be taken into account because they play a role when boiler efficiency between different types of boilers is compared. Therefore the system boundary for boiler efficiency loosely includes some but not all equipment in the boiler house.

A system boundary for boiler efficiency measurement is luckily usually determined in the applicable standard; i.e., the components belonging to the system boundary are defined in them exactly. All flow values are recorded when they cross the system boundary.

- Fuel handling, conveying, and feeding equipment are outside the boiler system boundary. Coal crushers and the associated equipment are within the system boundary as they are a part of an internal flow loop.
- Air fans and air ducts are outside the boiler system boundary. Equipment starting from the first heat transfer surface, the air preheater, is within the system boundary.
- The furnace with its associated equipment is within the boiler system boundary. Ash handling is outside the boiler system boundary.

- Flue gas cleaning equipment is outside the boiler system boundary. The fan for flue gas recirculation is outside the boiler system boundary.
- All steam and water heat exchangers that cool the flue gas are inside the boiler system boundary.
- Forced circulation pumps are inside the system boundary.
- The control system, instrumentation, and electrification are outside the boiler system boundary.

3.6.1 Useful Heat Output

Useful heat input includes all heats at to all steam flows. The value of course depends on the boiler type and the reason we are calculating the boiler efficiency. Normally the useful heat output can be defined as

$$Q_{abs} = Q_{ms} + Q_{rh} + Q_{bd} \qquad (3.2)$$

where

Q_{ms} is the heat transferred to main steam

Q_{rh} is the heat transferred to reheat steam

Q_{bd} is the heat transferred to blowdown

Usually no credit is given to steam used to heat air or for sootblow.

3.6.2 Heat and Energy Input

The energy input has two components. One is proportional to the fuel flow, and the other does not depend on the fuel flow. Energy flows that depend on the fuel flow are:
- chemical energy in the fuel, H_u (heat of combustion)
- energy included in the fuel preheating, Q_f
- energy included in the air preheating, Q_a

Examples of energy flows that are somewhat independent of the fuel flow are:
- shaft powers of the flue gas and air fans
- shaft powers of circulation pumps
- energy input by the flue gas recirculation fan.

It is customary to treat the useful heat input as a difference of input and output values. That is, useful heat is the difference between output flows and input flows. Therefore it is logical that those energy flows are not considered input flows. Such input flows are:
- heat in feedwater
- heat in desuperheating water flow
- Heat in incoming steam flow to the reheater.

3.6.3 Determining Efficiency With the Direct Method

When mass flows, specific heat values, and temperatures are known, the heat input with preheated air and fuel can be calculated using, e.g., the following simplified formula:

$$\eta = \frac{Q_{ms} + Q_{rh} + Q_{bd}}{H_f{}^* m_f + Q_f + Q_a + \sum P} \tag{3.3}$$

or

$$\eta = \frac{m_{ms}{}^*(h_{ms} - h_{fw}) + m_{rh}{}^*(h_{rh,out} - h_{rh,in}) + m_{bd}{}^*(h_{bd} - h_{fw})}{H_f{}^* m_f + m_f{}^*(h_{f,out} - h_{f,in}) + m_a{}^*(h_{a,out} - h_{a,in}) + \sum P} \tag{3.4}$$

where

H_f is the heating value of fuel
m_f is the fuel mass flow
Q_f is the heat transferred to preheated fuel
Q_a is the heat transferred to preheated air
$\sum P$ is the sum of mechanical and electrical energy input flows
m_{ms} is the main steam mass flow
m_{rh} is the reheat steam mass flow
m_{bd} is the blowdown mass flow
m_a is the air mass flow
h_{ms} is the enthalpy of main steam
h_{fw} is the enthalpy of feedwater steam
h_{bd} is the enthalpy of blowdown
$h_{rh,out}$ is the outlet enthalpy of reheating steam
$h_{rh,in}$ is the inlet enthalpy of reheating steam
$h_{f,out}$ is the outlet enthalpy of fuel
$h_{f,in}$ is the inlet enthalpy of fuel
$h_{a,out}$ is the outlet enthalpy of air
$h_{a,in}$ is the inlet enthalpy of air

The above formula is a very simple formula. It ignores most of the energy flows that cross the boundary. We note that at least the following streams have not been accounted for:

- condensate
- leakage air
- sootblowing
- other electricity flows
- flows to and from flue gas cleaning
- atomizing steam
- auxiliary fuels
- heating, ventilating, and air-conditioning streams
- heat stored in ash (solid or molten).

It can be argued that these streams are minor and do not affect the calculation. It is difficult to assess their effect a priori without measuring them. Therefore the need to measure many flows is one of the main problems in direct efficiency measurement.

Another problem arises from the theory of mathematical uncertainty associated with the measurements. As each and every stream needs to be measured, the error in the efficiency quickly becomes very large. Therefore the direct method is very seldom used in practice.

3.6.4 Determining Efficiency With the Indirect Method

The efficiency equation can be arranged also as (EN 12952-15, 2003)

$$\eta = 1 - \frac{Q_{in} - Q_{abs}}{Q_{in}} \qquad (3.5)$$

and even further as

$$\eta = 1 - \sum \frac{Q_{l,i}}{Q_{in}} \qquad (3.6)$$

where
   $Q_{l,i}$ is the i:th heat lost

The main losses in a steam generator are the following:
- Heat lost with the flue gases, which can be further divided as the sum of:
  - heat lost with the dry flue gases
  - heat lost with the water in the flue gases
- Losses of unburned combustible fuel, which can be further divided as the sum of:
  - losses in the incombustible in ash
  - losses in the incombustible in flue gases
- Sensible heat in the ashes
- Radiation and conduction losses.

Most of these losses can be estimated to a greater accuracy than the actual flows. The indirect method therefore gives higher accuracy when estimating the efficiency of the steam generator.

The biggest heat loss from a steam generator is the heat lost with the exiting flue gas. The flue gas loss depends on the final flue gas temperature and the amount of flue gases. Thus the higher the air ratio, the higher the flue gas losses.

The flue gases should leave the boiler at a temperature as low as possible to minimize the flue gas losses. Usually either economics, equipment, or corrosion issues limit the flue gas to $150-200°C$.

Neither all of the carbon nor all of the carbohydrates in the fuel fed into the boiler will combust. Reaction heat lost from incomplete reactions is termed incombustible. Actually, thermodynamics limit the progress of combustion reactions to some finite value. This means that there is always some CO, H_2, and other hydrocarbons present in the flue gases.

Total burning of solid fuels is difficult as char combustion is usually a slow process. This means that some carbon remains in the ash drawn from the furnace bottom. All solid biomass fuels contain some complex hydrocarbons. When they are burned, some these end up as fine unburned particles called soot. Soot is mostly unburned carbon residue and so lowers the boiler efficiency.

Heat is lost as the residue of combustion, hot ash, exits the furnace. To know the ash loss, the ash temperature, flow, and enthalpy must be determined. Fortunately the ash loss is usually small enough that even high inaccuracies in ash loss determination usually have a negligible effect on the accuracy of overall efficiency. If ash is exiting the furnace molten (e.g., in recovery boilers), then losses are significantly higher.

There are some losses from the hot boiler walls to the surroundings. These are called radiation and convection losses. Part of this energy increases the incoming air temperature. Usually radiation and convection losses are determined from a suitable diagram, Fig. 3.12.

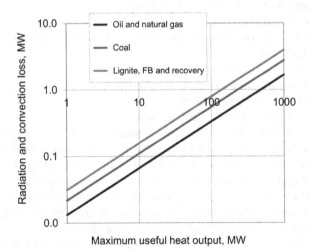

Figure 3.12 Radiation losses of boilers.

3.6.5 ASME PTC-4

In the United States and relevant countries the ASME PTC-4 performance test code is in use. It is drafted by the American Society of Mechanical Engineers (ASME, 2008). The main difference to EN 12952-15 is that as input the higher heating value of fuel is used. Therefore one needs to account as loss the latent heat of water exiting with the flue gas. Numerical values of efficiency derived from ASME PTC-4 are thus significantly lower than from EN 12952-15. In practice the core of the calculations is fairly similar.

3.6.6 DIN 1942

Very popular in its time was the DIN 1942 Acceptance Testing of Steam Generators (DIN 1942, 1990). It was drafted by the German Institute for Standardization. The contents are very close to the newer EN standard (EN 12952-12, 2003). DIN 1942 is currently only used to check older boiler performance for comparison of previous data.

3.6.7 Own Power Demand

A steam boiler needs pumping energy to be able to insert feedwater to produce pressurized steam. The higher the pressure, the higher the own power demand (Table 3.3).

Table 3.3 Own Power Demand

Pressure MPa	Own Power Percentage of Production	Own Power, kW/kg $_{steam}$/s	Pumping Power, kW/kg $_{steam}$/s
40	3.5	25	5
80	4.5	35	12
120	5.0	45	20
160	5.5	55	25
200	6.0	60	30
350	6.5	65	35

3.7 Placing of Heat Transfer Surfaces, Example

These calculations are for boiler surface heat in an example boiler, as seen in Fig. 3.13. The boiler main data is as follows:
- main steam flow 40.0 kg/s
- steam temperature 540°C
- steam pressure 13.0 MPa
- air preheating 30–>300°C
- economizer 200–>307°C
- flue gas flow 50 kg/s, fuel 2.5 kg/s
- flue gas outlet temperature 140°C
- superheater pressure loss 1 MPa
- blowdown 3%
- desuperheat, feedwater, 9.5%

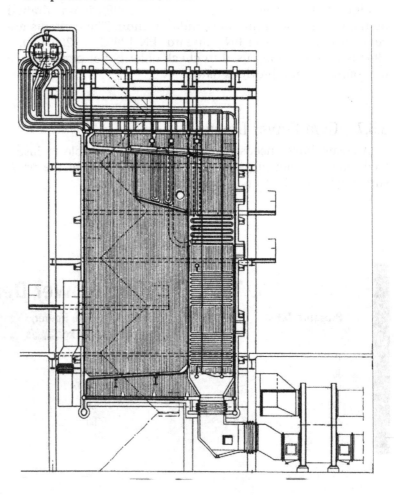

Figure 3.13 Side view of example boiler.

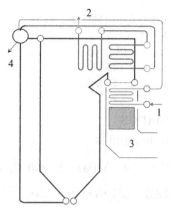

Figure 3.14 Simplified steam water system of example boiler 1-feedwater into the boiler, 2-main steam from the boilers, 3-flue gas from the boiler, 4-blowdown from steam drum.

Calculate heat surface loads. Compare real heat surface loads to ideal loads. Calculate heat to air heating, adiabatic temperature in the furnace, and furnace exit temperature.

From Fig. 3.13 we can deduce the steam water system. The deduced connection is shown in Fig. 3.14. Based on the steam water system, we can form the steam water mass flows from the following.

3.7.1 Mass Balance

$$q_{ms} - q_{dsh} + q_{bd} = q_{fw} \qquad (3.7)$$

Since we know from the main data that

$$q_{ms} = 40 \text{ kg}/s \quad \text{and} \quad q_{dsh} = 0.094*40.0 = 3.8 \text{ kg}/s \qquad (3.8)$$

then

$$q_{fw} = 40.0 - 3.8 + 1.2 = 37.4 \text{ kg}/s \qquad (3.9)$$

3.7.2 Enthalpies

Referring to Fig. 3.14.

Point 1   Feedwater in, temperature 200°C, pressure 14.5 MPa
Point 2   Main steam out, temperature 540°C, pressure 13.0 MPa
Point 3   Flue gas out, temperature 140°C
Point 4   Blowdown, saturated temperature, pressure 14.0 MPa

$$h \, (540°C, \, 13.0 \, MPa) = 3443.3 \, kJ/kg$$
$$h'' \, (14.0 \, MPa) \qquad = 2542.2 \, kJ/kg$$
$$h' \, (14.0 \, MPa) \qquad = 1570.9 \, kJ/kg$$
$$h \, (200°C, \, 14.5 \, MPa) = 857.9 \, kJ/kg$$
$$h \, (307°C, \, 14.0 \, MPa) = 1378.3 \, kJ/kg$$

3.7.3   Heat Demands

Economizer heat demand is

$$\Phi_{eco} = q_{fw} * (h(307°C, \, 14.0 \, MPa) - h \, (200°C, \, 14.5 \, MPa)) \lim_{x \to \infty}$$
$$= 37.4 * (1378.3 - 857.9) = 19.5 \, MW$$

$$(3.10)$$

Superheating heat demand is

$$\Phi_{sh} = (q_{ms} - q_{dsh}) * (h(540°C, 13.0 MPa) - h(14.0 MPa))$$
$$+ q_{dsh} * (h(540°C, 13.0 MPa) - h(200°C, 14.5 MPa))$$
$$= (40.0 - 3.8) * (3443.3 - 2542.2) + 3.8 * (3443.3 - 857.9) = 38.8 MW$$

$$(3.11)$$

Total heat demand is

$$\Phi_{tot} = q_{ms} * h(540°C, \, 13.0 \, MPa) + q_{bd} * h(14.0 \, MPa)$$
$$- (q_{fw} + q_{dsh}) * h(200°C, \, 14.5 \, MPa)$$
$$= 40 * 3443.3 + 1.2 * 1570.9 - (35 + 3.8) * 857.9 = 106.3 \, MW$$

$$(3.12)$$

Evaporative heat demand is

$$\Phi_{ev} = \Phi_{tot} - \Phi_{eco} - \Phi_{sh} = 106.3 - 19.5 - 38.8 = 48.0 \, MW \quad (3.13)$$

Ideally the superheating heat demand with no desuperheating is

$$\Phi_{sh} = q_{ms} * (h(540°C, \, 13.0 \, MPa) - h(14.0 \, MPa))$$
$$= 40.0 * (3443.3 - 2542.2) = 32.0 \, MW$$

$$(3.14)$$

Ideally the economizer heat demand with no blowdown and no margin is

$$\Phi_{eco} = q_{fw} * (h(14.0 \, MPa) - h(200°C, \, 14.5 \, MPa))$$
$$= 40 * (1570.9 - 857.9) = 28.5 \, MW$$

$$(3.15)$$

Ideally the evaporative heat demand with no blowdown and no desuperheating is

$$\Phi_{ev} = q_{ms} * (h''(14.0 \text{ MPa}) - h'(14.0 \text{ MPa}))$$
$$= 40 * (2542.2 - 1570.9) = 42.9 \text{ MW} \tag{3.16}$$

Ideally the total heat demand is

$$\Phi_{tot} = \Phi_{eco} + \Phi_{ev} + \Phi_{sh} = 32.0 + 28.5 + 42.9 = 103.4 \text{ MW} \quad (3.17)$$

In the ideal, simple cycle the total heat demand is 2.7% less. This is because heat is lost with the blowdown flow. There is significantly more superheating. Attemperating or desuperheating means that some of the preheating and evaporative heat flows are transferred to the superheating. There is more evaporation as the temperature from the economizer is less than the saturation temperature. There is significantly less preheating because of reduced flow and reduced temperature increase (Table 3.4).

Air preheating heat demand, $c_{pAir} = 1.0$ kJ/kgK

$$\Phi_{ah} = q_{air} * c_{p,air}(300°C - 30°C) = 47.5 * 1.0 * (300 - 30) = 12.8 \text{ MW} \tag{3.18}$$

Adiabatic combustion temperature, $c_{pFG} = 1.3$ kJ/kgK

$$\Phi_{tot} + \Phi_{ah} = q_{fg} * c_{p,fg}(T_{ad} - 140°C) \rightarrow$$
$$T_{ad} = (106.3 + 12.8)/(1.3 * 50) + 140°C = 1971°C \tag{3.19}$$

Furnace outlet temperature, assuming 95% of evaporation in furnace

$$T_{F,o} = 1971 - (48 * 0.95)/(1.3 * 50) = 1200°C \tag{3.20}$$

If a lower furnace outlet is required, then air preheating to a lower temperature is recommended.

Table 3.4 Comparison of Heats

	Ideal	Real	Difference
Economizing	28.5 MW	19.5 MW	−46.2%
	26.8%	18.3%	
Evaporation	42.9 MW	48.0 MW	10.6%
	40.4%	45.2%	
Superheating	32.0 MW	38.8 MW	17.5%
	30.1%	36.5%	
Total	103.4 MW	106.3 MW	2.7%
	97.3%	100%	

References

Advances in power station construction, 1986. Central Electricity Generating Board. Pergamon Press, Barnwood, Gloucester, UK, 759 p. ISBN 0080316778.

ASME, 2008. PTC-4 Fired Steam Generators. Performance Test Code 4:2008. American Society of Mechanical Engineers, New York, NY.

Bidard, R., Bonnin, J., 1979. Energetique et turbomachines. (Theory of energy and turbomachinery), Eyrolles, Paris, 742 p. ISSN 0399-4198 (in French).

DIN 1942, 1990. Acceptance testing of steam generators. Deutsches Institut für Normung e.V., Berlin, February 1994.

EN 12952-15, 2003. Water tube boilers and auxiliary installation—Part 15: acceptance test on steam generators. 12952-15:2003 E, 2003, CEN European Committee for Standardization.

Huhtinen, M., Kettunen, A., Nurminen, P., Pakkanen, H., 1999. Höyrykattilatekniikka. (Steam boiler technology). EDITA, Helsinki, 316 p. ISBN 951371327X (in Finnish).

Kjær, S., et al., 2002. The advanced supercritical 700°C pulverised coal-fired power plant. VGB Power Tech. 82 (7), 46–49.

Raiko, M., Gronfors, Tom, H.A., Haukka, P., 2003. Development and optimization of power plant concepts for local wet fuels. Biomass Bioenerg. 24 (1), 27–37.

Schmidt, E., 1989. Properties of Water and Steam in SI-Units: 0–800°C, 0–1000 bar. Springer, Berlin, p. 206. ISBN 3540096019.

Smith, J. W., 1998. Babcock & Wilcox Company supercritical (once through) boiler technology. Babcock & Wilcox Technical paper, BR-1658, Barberton, OH, 9 p.

Teir, S., 2004. Steam Boiler Technology. Energy Engineering and Environmental Protection publications, Helsinki University of Technology, Department of Mechanical Engineering, 2nd ed, p. 215. ISBN 9512267594.

4

STEAM—WATER
CIRCULATION DESIGN

CHAPTER OUTLINE

Steam Generation from Biomass. DOI: http://dx.doi.org/10.1016/B978-0-12-804389-9.00004-6
© 2017 Elsevier Inc. All rights reserved.

For a boiler to operate properly the steam—water circulation must be designed for large variations of load, manageable temperature differences in parallel tubes, and low possibility of erosion inside the tubes. The main areas in steam—water side circulation design are choosing the right type of circulation, the dimensioning of downcomers and risers, the dimensioning of superheaters, and the dimensioning of boiler banks. Full steam—water side design also includes boiler startup systems, continuous blowdown systems, and venting and draining systems.

4.1 Classification of Steam—Water Side

There are two main groups of steam—water circulation boilers, Fig. 4.1. The first group comprises the large-volume boilers. In these the heating evaporates steam inside a large volume of water. For example, a domestic kettle could be considered to be a large-volume type.

The second group of boilers is where boiling occurs inside a tube filled initially with water. For example, the coffee maker operates on this principle. Most large modern boilers belong to this group. In addition there is a group that could be called others or miscellaneous. They have features from both of the

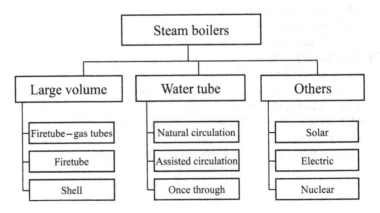

Figure 4.1 Classification of steam—water side.

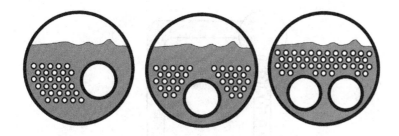

Figure 4.2 Fire tube—gas tube boiler designs. After Effenberger, H., 2000. Dampferzeugung (Steam boilers). Springer Verlag, Berlin, 852 p. ISBN 3540641750 (in German) (Effenberger, 2000).

aforementioned groups. Many of the nuclear plant steam circuits as well as solar power circuits belong to this mixed group. All groups, in spite of their design, are governed by the same laws.

4.1.1 Large-Volume Boilers

Large-volume boilers are usually fire tube boilers. Because of their design, they are limited by steam production (capacity) and operating pressure. The basic design has remained the same since the Scottish marine boilers of 1800.

In large-volume boilers the water circulates downwards at the edges of the boilers, Fig. 4.2. Steam bubbles rise, creating upward flow in the centerline of the boilers. The same mode of operation can be seen by watching a pan of water boiling on a stove.

4.1.2 Natural Circulation Boiler

Natural circulation is based on density differences. The same principle can be seen in, e.g., room on a cold winter's day. Warm air rises on the wall where the room is heated; subsequently, cooled, more dense air falls downwards on the opposite wall.

Natural circulation is caused by the density difference between saturated water and heated water partially filled with steam bubbles. In a natural circulation unit the water tubes are connected to a loop, Fig. 4.3. Heat is applied to one leg, called the raiser tube, where the water—steam mixture flows upward. Denser saturated water flows downward in an unheated leg called the downcomer.

4.1.3 Assisted Circulation

Assisted circulation is typical in heat recovery steam generator (HRSG) boilers and high-pressure units. Water from the

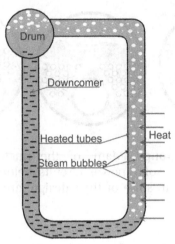

Figure 4.3 Principle of natural circulation.

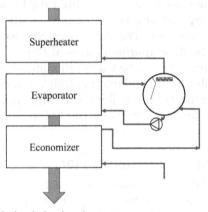

Figure 4.4 Assisted circulation in a heat recovery steam generator.

drum is pumped through evaporative surfaces. A synonym for assisted circulation is forced circulation (Fig. 4.4).

In controlled circulation a pump assists flow. Flow is regulated by orifices. This ensures even flow in all wall tubes. Controlled circulation is a trademark of ABB.

The most famous assisted circulation type is the La Mont boiler. The name comes from one manufacturer of these boilers. In La Mont boilers the drum pressure is usually below 19.0 MPa. The main advantage is that the designer can quite freely choose the tube pattern. Water flow through the tubes is controlled by appropriately sized orifices. The circulation ratio in La Mont boilers is from 4 to 10. Pressure loss in a circuit is usually from 0.1 to 0.3 MPa.

4.1.4   Once-Through Boiler

In once-through boilers the water flows continuously through the boiler, coming out as 100% steam at the main steam outlet. The circulation does not limit the pressure so the boiler can be built to withstand very high pressures. As positive circulation is retained with the pressure difference, the pressure loss through the boiler tends to be large. High pressure loss means high own power demand.

Feedwater purity must be very high as any contaminant tends to stay on the boiler walls. The feedwater must have about the same purity as the steam. The steam purity is dictated by the turbine requirements. Starting and shutting the once-through boiler is problematic. Because of the pumps, the mass flow densities in once-through boilers are high, Table 4.1.

The furnace tube arrangement is difficult. In natural and assisted circulation the furnace can be made of straight tubes of the same temperature. Once-through boilers employ various tube patterns to cover the furnace walls. Based on these types of circuit, the once-through boilers are divided into Sultzer Monotube boilers, Benson boilers, and Ramzin boilers (Fig. 4.5).

One of the first successful once-through boilers was the Sultzer Monotube boiler. Its trademark is the use of a bottle, where remaining water droplets are separated from the steam. The flow in the furnace uses U-shaped tube bundles.

A competitor to the Sultzer design, the Benson boiler had a very similar tube pattern (Fig. 4.6). Very rarely was the bottle or similar device used in a Benson boiler. To facilitate design the Benson boiler uses straight heated up-flow parts and unheated down-flow parts. In the Benson boiler there is no definite point where the evaporation ends and the superheating begins, when operated under the critical pressure.

Dividing each pass into several segments helps to maintain low temperature differences between adjacent tubes. Heat flux to tubes can vary because of localized fouling and tube placement.

Table 4.1 Mass Flow Densities

Surface	Mass Flow Density (kg/m²s)
Convective superheater	1000
Furnace tubes	2000—3000
Economizer	600

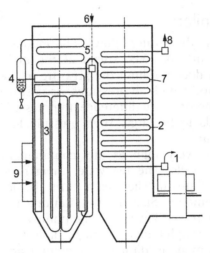

Figure 4.5 Sultzer Monotube circulation: 1, feedwater inlet; 2, economizing surface; 3, furnace tubes; 4, bottle; 5, superheating surface; 6, desuperheating spray; 7, superheating; 8, main steam out; 9, fuel input. From Doležal, R., 1967. Large Boiler Furnaces. Elsevier Publishing Company, 394 p. (Doležal, 1967).

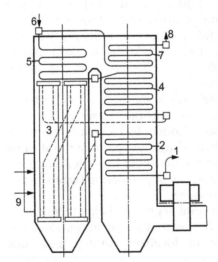

Figure 4.6 Benson boiler circulation: 1, feedwater inlet; 2, economizing surface; 3, furnace tubes; 4, backpass surface; 5, superheating surface; 6, desuperheating spray; 7, superheating; 8, main steam out; 9, fuel input. From Doležal, R., 1967. Large Boiler Furnaces. Elsevier Publishing Company, 394 p.

The Ramzin boiler was developed in Russia (Fig. 4.7). It uses much the same operating principles as the Sultzer Monotube boiler. The main distinctive feature of the Ramzin boiler is that the tubes circle the furnace. It is expensive to manufacture. Separation (4), similar to the Sultzer design, was added later. The main use of Ramzin boilers has been in the former Eastern Bloc.

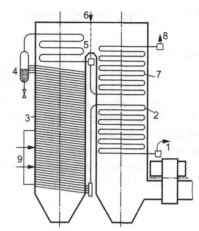

Figure 4.7 Ramzin circulation: 1, feedwater inlet; 2, economizing surface; 3, furnace tubes; 4, bottle; 5, superheating surface; 6, desuperheating spray; 7, superheating; 8, main steam out; 9, fuel input. From Doležal, R., 1967. Large Boiler Furnaces. Elsevier Publishing Company, 394 p.

In once-through boilers the evaporation is brought to completion. Therefore parts of the tubes must be operated with a very high vapor content. This kind of operation stresses certain tubes when water is sometimes separated from the tube walls and sometimes not. One helpful solution is to make parts of the tubes from rifled or internally finned tubes. The water—steam solution is brought into rotative motion and heavier water tends to stay longer at the tube walls. Spirally finned tubing increases wall wetting, decreases the possibility of departure from nucleate boiling (DNB) and is more expensive than straight tubing.

Because of the possibility of heat transfer instabilities at low mass flows, one needs to improve flow in low loads. Then a pump can be used for partial load operation. A part of the steam—water solution can be recirculated to the drum and pumped again through the boiler.

4.2 External Pipes

Most of the parts of the pressure vessel belong to the heat exchanger surfaces and circulation system. There is a system of pipes, equipment, and vessels that handles the feedwater. There is also a system of pipes that handles main steam. In addition there are pipes that belong to the desuperheating system, blowdown system, air preheating system, draining system, and air venting system. Finally there is piping assisting the operation of the safety valves. All these tubes and pipes are called external pipes.

4.2.1   Feedwater System

The feedwater system consists of piping, tanks, and equipment that pumps water to the inlet of the economizer, Fig. 4.8. Return condensate with additional demineralized water is pumped to the feedwater tank. There are always some steam and water losses that circulate. These losses must be covered with make-up water. Before it reaches the feedwater tank, the feedwater is preheated to improve steam generation efficiency. The feedwater tank has a storage capacity corresponding to 15—45 minutes of feedwater usage at maximum continuous rating (MCR). The feedwater tank ensures that there is enough

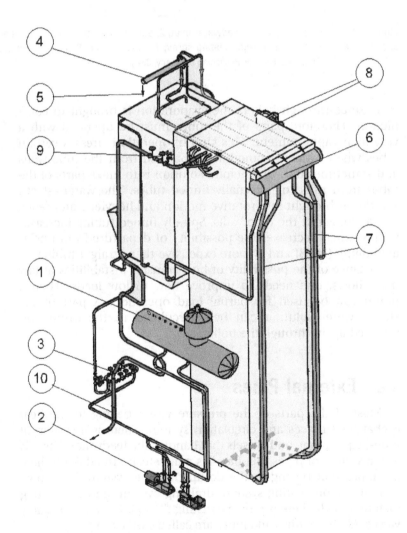

Figure 4.8 External piping: 1, feedwater tank; 2, feedwater pumps; 3, feedwater valves; 4, Dolezahl condensor; 5 spray water out; 6, steam drum; 7, downcomers; 8, superheater connectors; 9, main steam valves; 10, main steam pipe.

feedwater at all times. In addition to storage, deaeration is carried out in the incoming water. The purpose is to remove all traces of gases, especially oxygen, from the water.

Feedwater is pumped from the feedwater tank to the economizers. Typically the feedwater pumps are driven by electric motors. To ensure safety during pump failure there is usually more than one pump. Often one feeds these pumps from different transformers for redundancy reasons. Feedwater flow is controlled by feedwater valves and the pump speed.

Often with utility or large boilers there are feedwater preheaters with connecting steam lines and condensate lines.

4.2.2 Superheater Connecting Tubes

Each superheater must be connected to the following superheater. Typically the connection is made from both ends of the inlet and outlet headers. Superheater control is easier if the right side outlet flow is connected to the left side inlet flow, and the left side outlet flow is connected to the right side inlet flow (i.e., the connecting tubes are crossed).

Typically, superheater connection tubes also have desuperheating connections. In desuperheating, feedwater is sprayed to cool steam flow and thus to control steam temperature.

4.2.3 Feedwater Tank

The feedwater tank usually sits in the middle of the boiler room. It is situated at a height of 10–30 m from the floor so that the feedwater pumps can have the necessary room to operate.

The primary function of a feedwater tank is water storage. The feedwater flow is not constant but fluctuates, especially if the boiler has disturbances. Therefore the tank's design often allows for a water reserve of around 20 minutes to 1 hour, which is calculated from MCR feedwater consumption. When calculating feedwater volume height, about 500–600 mm from the bottom of the tank needs to be omitted, as a water level below this would lead to steam entering into the pumps with the feedwater. Similarly, to keep the gas separation operating, 400–600 mm from the top cannot be used. The feedwater tank is often made of carbon steel, typically P265GH (EN 10028:2).

The feedwater tank is also used to separate gases from the feedwater. This is done by steam stripping incoming water in the deaerator.

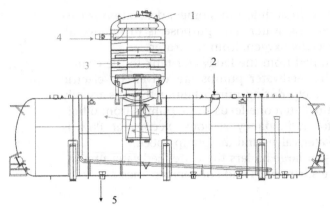

Figure 4.9 Feedwater tank: 1, deaerator; 2, steam in; 3, trays; 4 water in; 5, water out.

In Fig. 4.9 shows a conventional deaerator placed on top of the feedwater tank. Some manufacturers place their deaerator inside the tank, so in contrast to Fig. 4.9 no visible deaerator on top of the tank would be seen.

4.2.4   Feedwater Pump

Feedwater pumps are commonly centrifugal pumps. To ensure a high enough head, several (even dozens) of pumps have been arranged in series. A feedwater pump is most often driven by an electric motor for energy efficiency. Often steam turbine-driven feedwater pumps are used.

To ensure adequate feedwater flow a boiler must have at all times a slightly higher capacity for feedwater than for steam generation. Feedwater pumps can be common to several boilers. For safety, one needs enough feedwater pumping capacity even if the largest feedwater pump fails.

4.2.5   Dolezahl—Attemperator

The Dolezahl attemperator is used especially in industrial boilers, where feedwater quality is lower than in utility boilers. In a Dolezahl the feedwater condenses the steam. This condensate is sprayed using the static head formed to attemperate the produced steam. Normally saturated steam is used. High quality spray water is ensured as saturated steam has much higher purity than the corresponding feedwater (Fig. 4.10).

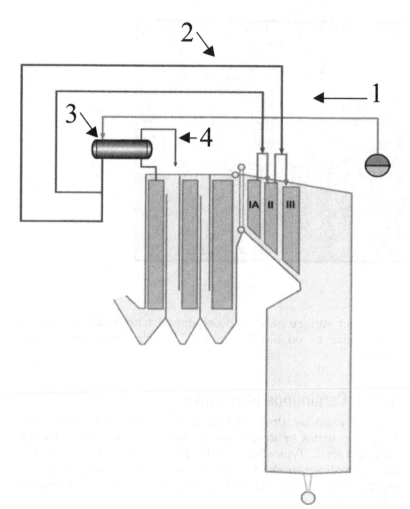

Figure 4.10 Dolezahl—attemperator: 1, steam flows from drum; 2, condensation in heat exchanger; 3, condensate sprayed to desuperheat; 4, heat transferred to feedwater.

4.2.6   Spray Water Group

In a spray water group the feedwater is injected into the steam. This increases the steam mass flow and decreases the steam temperature. Typically, spraying to different tubes is arranged close to one another for maintenance purposes. Also, the automatic spray water valve is usually doubled or tripled with hand valves. Often, because of location airing and emptying, valves need to be installed (Fig. 4.11).

4.2.7   Drainage and Air Removal

When the boiler is brought into operation, it must be filled with water and the air removed. All top points of each heat

Figure 4.11 Spray water group.

exchanger surface need to be equipped with small lines and a hand valve to enable the removal of air from the pressure vessel.

4.2.8 Continuous Blowdown

High-pressure drum water is removed by blowdown to keep impurities in steam—water circulation below the recommended level. Typically, blowdown is flashed to the continuous blowdown tank. Generated steam is conveyed to, for example, the feedwater tank to increase electricity-generating efficiency.

4.2.9 Condensate Collection

Cleaning water to the required high purity level is costly. Therefore all steam condensate flows from within the boiler house are collected in a separate condensate tank. Condensate is formed in, e.g., steam air preheaters. Collected condensate is pumped directly to the feedwater tank.

4.3 Principle of Natural Circulation

Most modern water tube boilers are natural circulation boilers. Natural circulation means that the steam water movement in the evaporative tubes is achieved without the use of external prime energy and mechanical means.

4.3.1 Driving Force

The driving force is the static head difference between water in downcomers and the steam—water mixture in the furnace tubes.

$$\Delta P_1 = (\rho_w - \rho_{mix})gh \qquad (4.1)$$

where

ΔP_1 is the flow losses in the circulation

ρ_w is the density of the water in the downcomers

ρ_{mix} is the density of the water—steam mixture in the heated section

h is the height of the circulation

An increase in pressure decreases the driving force. Generally the final steam content is kept well below 10%. This means that circulation ratios (the inverse of steam content as a mass ratio) of 15—25 are typically used. This also means that one needs to design the circulation to at least one magnitude higher flow than the maximum design steam flow.

4.3.2 Components of Natural Circulation

Downcomers are tubes that start from the steam drum and supply water to the lowest points of the furnace walls and boiler banks. Downcomers can be divided into main downcomers and connecting divider tubes (Reznikov and Lipov, 1985). It is customary for 2—5 large tubes to go straight down from the drum. From these tubes smaller divider tubes take water to the wall headers (Fig. 4.12).

Typically each heat transfer surface starts and ends with a header. Headers are larger tubes that connect all parallel tubes of a heat transfer surface. Their role is to divide the incoming liquid evenly to all tubes. They also try to collect the fluid without causing uneven flow.

Headers used to have various shapes. Even a square cross-section was used. Currently only tubular headers are used in industrial equipment. The flow velocity in headers should be kept at a minimum to make flows equal. A typical hand rule is to design the header larger than the sum of the flow areas that lead from it.

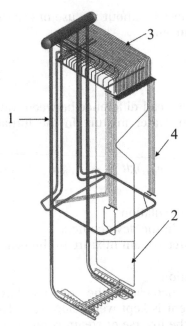

Figure 4.12 Natural circulation boiler: 1, water flows down; 2, steam water flows up water walls; 3, steam water mixture flows back with risers; 4, other surfaces (boiler bank) can be equally connected.

From the evaporating wall tube outlet headers the water–steam mixture is led with raisers into the cyclone separators, where the steam is separated from the water–steam mixture and the exiting steam is further dried in demisters. Separated water is led back to the downcomers.

4.3.3 Optimization of Natural Circulation Design

Natural circulation design is affected by several design factors. If we increase the furnace height, the driving head increases. We must try to avoid steam in downcomers. Efficient steam separators reduce the fraction of steam inside the drum. Inserting feedwater into the drum at a subsaturated state cools down and collapses the remaining bubbles. Minimizing the axial flow inside the drum helps to create equal flow in all parts of the boiler.

Natural circulation is also improved by higher heat flux in the lower part of the tubes. Therefore there needs to be careful consideration if non-typical furnace designs are used.

4.4 Steam Drum

The steam drum consists of a circular section welded from bent steel plate with two forged ends. The circular section has a number of inlets and outlets. A drum can be made from a single thickness sheet if the pressure is low, or from plates of multiple thickness if the pressure is high. One reason for different thicknesses is that sections with openings need to be thicker than sections without openings (Fig. 4.13).

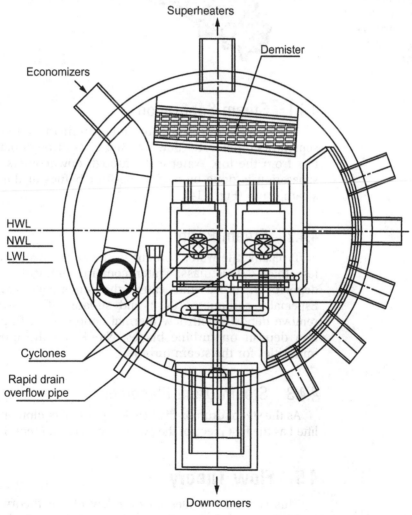

Figure 4.13 Steam drum.

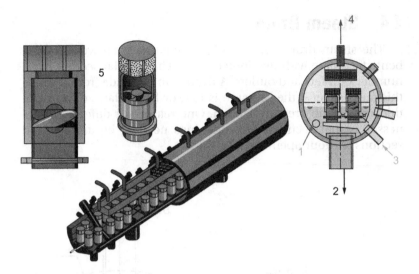

Figure 4.14 Placement of water separators in a drum: 1, feedwater in; 2, downcomers; 3, raisers; 4, steam out; 5, droplet separator.

4.4.1 Steam Separation

The steam—water mixture enters from the raisers. Cyclone separators force water down. Steam exits through baffle separators from the top. Water exits through downcomers. Additional separation is done by a wire mesh or baffles at the top of the steam drum (Fig. 4.14).

4.4.2 Steam Purity

Typically, the water content in the steam needs to be much less than 0.01% mass. Water contains impurities. Impurities deposit on tubes and cause the superheater to overheat. Especially carbonate (CO_3) and sulfate (SO_4) form hard-to-remove deposits with low thermal conductivity. Impurities will also deposit on turbine blades. Na+K are the most harmful impurities for the steam turbine.

4.4.3 Steam Drum Placement

As the steam drum is the heaviest piece of equipment, it is often lifted as the first piece of the pressure part equipment (Fig. 4.15).

4.5 Flow Theory

This section provides an overview of the theory needed to understand the design of steam—water tubes, calculate pressure losses and look at some possible operating problems.

Figure 4.15 Erection of steam drum under wintery conditions.

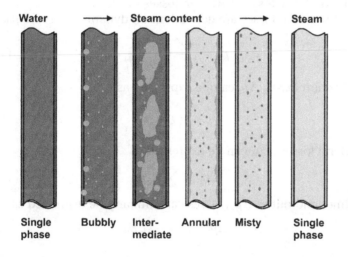

Figure 4.16 Types of flow with increasing steam content from left to right.

4.5.1 Types of Flow

The type of flow in a steam generator tube depends on the steam content (Collier, 1983). In Fig. 4.16 some of the possible flow configurations in a vertical tube are shown. The flow on the far left is undersaturated flow, where no steam bubbles exist. If heat is applied to saturated water then steam bubbles start to appear. In Fig. 4.16 the second tube from the left is heated and steam bubbles are generated at the tube surface and then transferred to the flow. This is the preferred nucleate boiling.

Two-phase flow (flow where, e.g., steam and water coexist) can be stable or unstable. A targeted stable flow pattern is shown in the third tube from the left in Fig. 4.16. Large steam bubbles are formed in the middle of the tube, but all the walls remain wetted. A larger amount of steam can lead to annular flow, where the whole middle of the tube is open to the steam flow. Finally, when the tube's steam content is very high, the tube walls will be dry and water will only exist as droplets or mist in the flow.

4.5.2 Pressure Loss in Single-Phase Flow

The tube side pressure drop is the sum of individual pressure drops

$$\Delta p_{ts} = \Delta p_b + \Delta p_i + \Delta p_o + \Delta p_f \tag{4.2}$$

Pressure losses = bend pressure losses + inlet pressure losses + outlet pressure losses + friction losses

All pressure losses are dependent on dynamic pressure p_d,

$$p_d = \frac{\rho w^2}{2} = \frac{8 q_m^2}{\rho \pi^2 d_i^4} \tag{4.3}$$

Friction losses Δp_f can be expressed as

$$\Delta p_f = \xi \frac{L}{d_i} p_d \tag{4.4}$$

Bend losses Δp_b can be expressed as

$$\Delta p_b = \sum \zeta_b p_d \tag{4.5}$$

Inlet and outlet Δp_{io} losses can similarly be expressed as

$$\Delta p_i + \Delta p_o = \zeta_{io} p_d \tag{4.6}$$

$$\Delta p_{io} = \zeta_{io} p_d \tag{4.7}$$

where
   L is the length of the tubes
   d_i is the inside diameter of the tubes
   ξ is the friction coefficient
   ζ is the loss coefficient for bend, inlet and outlet losses
   Pressure loss is a function of the dynamic head, tube length, and the inverse of the tube diameter.

4.5.3 Friction Loss Coefficient

In general the friction loss coefficient depends on the flow conditions. The pressure loss is different for different types of flow. The friction loss coefficient ξ for laminar flow is expressed by the Hagen—Poiseuille equation

$$\xi = \frac{64}{\text{Re}} \tag{4.8}$$

For turbulent flow in hydraulically smooth tubes, when Re is from 3000 to 100,000, the friction loss coefficient ξ is expressed by the Blasius equation

$$\xi = \frac{0.3164}{\sqrt[4]{\text{Re}}} \tag{4.9}$$

And for turbulent flow in rough tubes the pressure loss coefficient ξ is expressed by, e.g., Colebrook equation (Colebrook, 1939) (Fig. 4.17).

$$\frac{1}{\sqrt{\xi}} = 1.74 - 2\log\left(\frac{2k}{d_i} + \frac{18.7}{\text{Re}\sqrt{\xi}}\right) \tag{4.10}$$

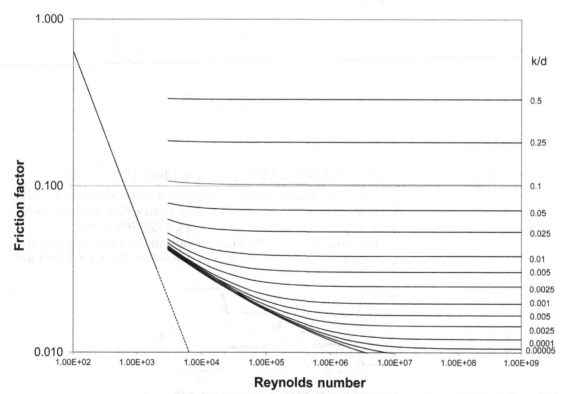

Figure 4.17 Moody chart for the friction factor of fully developed pipe flow for various k/d ratios as shown in right. After Moody, L.F., 1944. Friction factors for pipe flow. Trans. ASME 66 (8), 671—684 (Moody, 1944).

Usually these equations are shown as the Moody chart.

The pressure loss coefficient is lower for high Reynolds numbers. On the other hand, pressure loss is higher for high velocities and high Reynolds numbers because dynamic pressure increases as a square of velocity. The pressure loss coefficient is lower for large diameter tubes. Roughness, k, for drawn steel tubes is 0.025–0.05 mm. Therefore the pressure loss coefficient is usually close to 0.05 (Nikuradse, 1933). It should be noted that there is a transition region from smooth to rough surfaces. Flow behaves smoothly at the transition region.

Typical economizer flows in boilers have high Reynolds numbers, but superheater flows can be in the transition region from laminar to turbulent flow. It is therefore suggested that a combination equation be used to calculate the pressure loss. One such equation has been developed by Churchill (1972).

$$\xi_l = 8\left[\left(\frac{8}{\text{Re}}\right)^{12} + (A+B)^{-1.5}\right]^{1/12} \tag{4.11}$$

where

$$A = \left(2.457\ln\left[\left(\frac{7}{\text{Re}}\right)^{0.9} + 0.27\frac{k}{d_i}\right]\right)^{16}$$

$$B = \left(\frac{37530}{\text{Re}}\right)^{16}$$

4.5.4   Pressure Loss in Tube Inlet and Outlet

The area ratio for sudden contraction and widening, Fig. 4.18, is defined as A_1/A_2. For sudden flow contraction the loss coefficient depends on the area ratio, (Kays, 1950) (Fig. 4.19).

For normal cases the flow at the inlet is turbulent and the area ratio A_1/A_2 approaches zero. Therefore, the inlet losses are

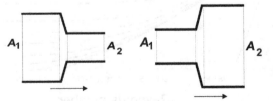

Figure 4.18 Two normal types of fittings: left flow contraction; right flow widening.

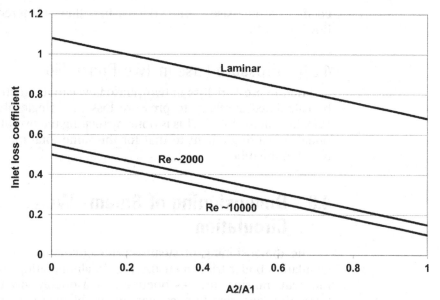

Figure 4.19 Loss coefficient with flow contraction. After Kays, W.M., 1950. Transactions of ASME. 72, 1067–1074.

about 0.5. For sudden flow widening we can assume that all kinetic energy is dissipated. For outlets the pressure loss coefficient can be given as

$$\zeta_o = \left(1 - \frac{A_1}{A_2}\right)^2 \qquad (4.12)$$

This outlet flow loss coefficient is thus for all practical purposes close to 1.0. For steam generator conditions the inlet and outlet loss coefficient ζ_{io} is approximately 1.5.

4.5.5 Pressure Loss in Bends

To calculate bend losses the method of Hooper (1984) can be used

$$\zeta_b = \frac{1}{Re} + b_b\left(1 + \frac{25.4}{d_i}\right) \qquad (4.13)$$

where the bend tightness coefficient b_b depends on the tube mean bend radius. For typical steam generator tubes

$$b_b = 0.34 - 0.02r_b; \; r_b < 2$$
$$b_b = 0.90 - 0.30r_b; \; r_b \geq 2 \qquad (4.14)$$

It must be noted that bend losses are strongly influenced by the appropriate Reynolds number and so the bend losses coefficient increases with decreasing flow (Kitteredge and Rowley, 1974).

4.5.6 Pressure Loss in Two-Phase Flow

Friction pressure loss in two-phase flow can in practical cases be calculated similarly to pressure loss in a single-phase flow (VDI heat atlas, 1993). This is done by relating the two phase frictional pressure gradient to that for the same total mass flow of liquid in the tube.

4.6 Dimensioning of Steam—Water Circulation

The dimensioning of steam—water circulation in a natural circulation boiler tries to ensure that in all operating conditions the heat transfer surfaces operate at reasonably low temperatures, thus ensuring high availability. Problems in steam—water circulation will cause high local temperatures and lead to fishmouth-type tube failures.

In a natural circulation boiler the density difference between the subcooled water in the downcomers and the water—steam mixture in the riser pipes causes the water to flow through the evaporator surfaces. Careful design of supply and riser pipes and inlet and outlet headers ensures good water recirculation in all parts of the circuitry — also during load variations.

4.6.1 Downcomers and Dividers

Downcomers are tubes that start from the drum and lead water to furnace walls and boiler banks, see (Fig. 4.12). Downcomers can be divided to main downcomers and connecting tubes. It is customary that 2—5 large tubes go straight down from the drum. Often in the end of downcomers is a connecting header. From these smaller divider tubes take water to the wall headers. Wall headers are large horizontal tubes. Wall tubes start from lower wall headers and end to upper wall headers. It is important to try to keep the flow velocity in the downcomers low to avoid problems in circulation design. Often velocities between 4 and 6 m/s are chosen. Extra care must be taken to ensure that there is equal pressure at the lowest dividing header.

Dividing header and headers of each wall are connected with divider tubes. One must use plenty of divider tubes to ensure that each wall header receives enough circulating water.

4.6.2 Wall Tubes

Wall tubes are the membrane wall tubes. The aim of the circulation is that in each tube an equal amount of water is circulating. Typically this cannot be achieved as some tubes receive more heat than others. Similarly some tubes are longer or contain more bends and thus have higher flow losses than others. So it is typical that flows in some tubes can differ up to 30% from the average flow.

It is important that flow velocity in the wall tubes is initially low as the steam generation increases volume and thus flow speed. Generally inlet velocities between 0.3 and 1.0 m/s are chosen for maximum steam generation.

4.6.3 Raisers

Raisers are tubes that start from the upper headers of the furnace walls and other evaporating surfaces and connect them to the steam drum. Main raisers are those from each furnace wall to the drum. It is important that wall raisers are placed not too far apart. A large number of raisers means small flow differences between different parts of the walls.

Raisers are expensive as there must be a corresponding hole in the steam drum. The more holes we need, the thicker the steam drum wall thickness becomes. It is typical to choose flow velocities between 8 and 12 m/s for maximum steam generation.

4.6.4 Feedwater Pump Head

The feedwater pump must produce enough overpressure (pump head) to be able to maintain the drum level in the boiler in all conditions.

$$\Delta p = p_p - p_i + \Delta p_f + \Delta p_v \qquad (4.15)$$

where
  Δp is the total required pump head, MPa
  p_p is the maximum operating pressure at the drum, MPa
  Δp_f is the loss in the feedwater piping and economizer, MPa
  Δp_v is the loss in the feedwater flow control valve, MPa
  ρgh is the pressure required to overcome the height difference between feedwater tank lower level and drum level, MPa
  usually

$$p_p - p_i = \rho gh \qquad (4.16)$$

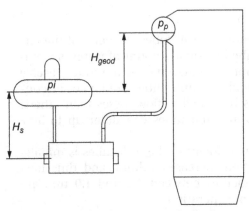

Figure 4.20 Determination of the feedwater pump head; p_p pressure in the drum; p_i pressure in the feedwater tank; H_s height difference between pump inlet and water level in feedwater tank; H_{geod} height difference between water levels in feedwater tank and drum.

It should be remembered that the feedwater pump must be placed well below the feedwater tank (Fig. 4.20). This ensures additional pressure at pump inlet called nominal suction pressure head (NSPH). In the feedwater tank the water is at saturated pressure. When water flows to the feedwater pump there will be pressure losses. At the pump impeller the change in velocity produces static pressure decline. If suction pressure is not over corresponding saturated pressure, then cavitation will occur and the pump impeller will be destroyed in a relatively short timespan.

4.7   Calculation of Superheater Pressure Loss

To illustrate the use of simple flow theory to calculate complex flows the superheater case is presented as an example. Calculation of superheater pressure losses is iterative as there is more than one parallel tube of different length. Let us assume that the estimated conditions are

$$T_{av} = \frac{T_{in} + T_{out}}{2} = \frac{285°C + 315°C}{2} = 300°C \text{ and}$$

$$p_{av} = \frac{p_{in} + p_{out}}{2} = \frac{(69.0 + 68.44)bar}{2} = 68.72 \ bar$$

Then the average density of steam is 33.097 kg/m^3 and dynamic viscosity is 1.98×10^{-5} kg/sm based on steam–water tables. We can first estimate an equal mass flow through every tube

$$q'_{m1} = \frac{q_{m,\text{tot}}}{n_{\text{pipe}} \times n_{\text{plate}}} = \frac{22.7}{17 \times 5} \text{ kg/s} = 0.267 \text{ kg/s}$$

where

$q_{m,\text{tot}}$ is the total mass flow through the whole superheater, kg/s

n_{pipe} is the number of parallel tubes in a superheater platen

n_{plate} is the number of platens in a superheater

Flow velocity w'_1 in a tube is then

$$w'_1 = \frac{q'_{m1}}{0.25\rho\pi d_i^2} = \frac{0.267}{0.25 \times 33.097 \times \pi \times 0.0343^2} \text{ m/s} = 8.73 \text{ m/s}$$

where

ρ is the flow density, kg/m^3

d_i tube inside diameter, m

We can then calculate Re'_1 based on velocity w'_1

$$\text{Re}'_1 = \frac{8.73 \times 0.0343 \times 33.097}{1.98 \times 10^{-5}} = 500532$$

and calculate the friction factor based on Re'_1

$$\xi'_1 = \frac{0.25}{\left(\log\left(\frac{0.03}{3.7 \times 34.3} + \frac{5.74}{500532^{0.9}}\right)\right)^2} = 0.01979$$

After calculating the additional losses from the tube bends we can sum all additional losses $\Sigma\zeta = 4.98$.

Total pressure loss through the superheater is

$$\Delta p'_1 = \left(\frac{40.2}{0.0343} \times 0.01979 + 4.98\right) \times \frac{33.097 \times 8.73^2}{2} \text{Pa} = 35.533 \text{ kPa}$$

We can calculate the flow velocity second time w'_2 assuming the calculated pressure loss and the calculated friction factor.

$$w'_2 = \frac{2 \times \Delta p'_1}{\rho \times \left(\frac{L_1}{d_i} \times \xi'_1 + \zeta_2\right)} = \sqrt{\frac{2 \times 35.533 \times 10^3}{33.097 \times (23.19 + 7.96)}} \text{ m/s} = 8.30 \text{ m/s}$$

Recalculate Re'_2 from w'_2

$$\text{Re}'_2 = \frac{8.30 \times 0.0343 \times 33.097}{1.98 \times 10^{-5}} = 475878$$

Recalculate friction losses from Re'_2

$$\xi'_2 = \frac{0.25}{\left(\log\left(\frac{0.03}{3.7 \times 34.3} + \frac{5.74}{475878^{0.9}}\right)\right)^2} = 0.01982$$

Let us recalculate w'_1 once more assuming that the pressure difference equals the pressure difference in tube 1, but using w'_2 as a friction factor.

$$w'_2 = \sqrt{\frac{2 \times \Delta p'_1}{\rho \times \left(\frac{L_2}{d_i} \times \xi'_2 + \zeta_2\right)}} = \sqrt{\frac{2 \times 35.533 \times 10^3}{33.097 \times (42.7 + 7.96)}} \text{ m/s} = 6.51 \text{ m/s}$$

Let us sum all mass flows together for $q'_{m,plate}$

$$q'_{m,plate} = \sum_{n=1}^{5} q'_{m,n} = 1.060 \text{ kg/s}$$

Calculate the mass flows through each tube

$$q_{m,1} = \frac{q'_{m,1}}{q'_{m,plate}} q_{m,plate} = \frac{0.267}{1.06} \times 1.335 = 0.336 \text{ kg/s}$$

where $q'_{m,1}$ is the mass flow corresponding to the mass flow at the Table 4.2 $q'_{m,plate}$

$$q_{m,plate} = \frac{q_{m,tot}}{n_{plate}} = \frac{22.7}{17} \text{ kg/s} = 1.335 \text{ kg/s}$$

Table 4.2 Mass Flows q'_m for All Tubes

	q'_m kg/s	w' m/s	Re' —	$L/d \times \xi'$ —	Δp kPa
Tube 1	0.267	8.73	501623	23.19	35.55025
Tube 2	0.199	6.51	477001	42.70	
Tube 3	0.198	6.48	473239	42.72	
Tube 4	0.198	6.48	473239	42.72	
Tube 5	0.198	6.48	473239	42.72	

Table 4.3 Accurate Mass Flows

	Flow/pipe (kg/s)
Tube 1	0.336
Tube 2	0.251
Tube 3	0.249
Tube 4	0.249
Tube 5	0.249

The accurate mass flows for all tubes have been calculated to Table 4.3.

When real mass flows are known, then the velocities can be recalculated, e.g.,

$$w_1 = \frac{q_{m1}}{0.25\rho\pi d_i^2} = \frac{0.336}{0.25 \times 33.097 \times \pi \times 0.0343^2}\, \text{m/s} = 10.99 \text{ m/s}$$

And the Reynolds number can be recalculated, e.g.,

$$\text{Re}_1 = \frac{10.99 \times 0.0343 \times 33.097}{1.98 \times 10^{-5}} = 630108$$

And the new friction factor can be recalculated, e.g.,

$$\xi_1 = \frac{0.25}{\left(\log\left(\frac{0.03}{3.7 \times 34.3} + \frac{5.74}{630108^{0.9}}\right)\right)^2} = 0.01965$$

and the accurate pressure loss calculated

$$\Delta p_1 = \left(\xi_1 \frac{L_1}{d_{i,1}} + \Sigma\zeta_1\right)\frac{\rho w_1^2}{2}$$

$$= \left(0.01965 \times \frac{40.2}{0.0343} + 4.98\right) \times \frac{33.097 \times 10.99^2}{2}\, \text{Pa}$$

$$= 55.984 \text{ kPa}$$

Final pressure losses and Reynolds numbers and velocities are shown in Table 4.4.

Table 4.4 Final Pressure Losses

	w m/s	Re –	$\xi^* L/d$ –	$\sum \zeta$ –	ΔP bar
Tube 1	10.99	631539	23.03	4.98	0.560
Tube 2	8.20	470904	42.72	7.96	0.564
Tube 3	8.16	468556	42.73	8.46	0.564
Tube 4	8.16	468556	42.73	8.46	0.564
Tube 5	8.16	468556	42.73	8.46	0.564

References

Churchill, S.W., 1972. Friction factor equation spans all fluid regimes. Chem. Eng. 84 (24), 91–92.

Colebrook, C.F., 1939. Turbulent flow in pipes with particular reference to the transitions region between the smooth and rough pipe laws. J. Inst. Civil Eng. 11 (1938/39), 133–156.

Collier, J.G., 1983. Boiling and evaporation. In: Heat Exchanger Design Handbook, HEDH, Part 2, Fluid Mechanics and Heat Transfer. VDI, Düsseldorf, pp. 2.7.1–2.7.4-12. ISBN 318419082X.

Doležal, R., 1967. Large Boiler Furnaces. Elsevier Publishing Company, 394 p.

Effenberger, H., 2000. Dampferzeugung (Steam boilers). Springer Verlag, Berlin, 852 p. ISBN 3540641750 (in German).

Hooper, W.B., 1984. The two-K method predicts head losses in pipes and fittings. Chem. Eng. 88 (17), 96–100.

Kays, W.M., 1950. Trans. ASME. 72, 1067–1074.

Khalil, E.E., 1990. Power Plant Design. Gordon & Breach, 370 p. ISBN 0856265055.

Kittredge, C.P., Rowley, D.S., 1974. Resistance coefficients for laminar and turbulent flow through ½ inch valves and fittings. Transactions of ASME. J. Eng. Power. 1579–1586.

Moody, L.F., 1944. Friction factors for pipe flow. Trans. ASME. 66 (8), 671–684.

Nikuradse, J., 1933. Strömungsgesetz in rauchen rohren (Flow loss in rough tubes). Forshungsheft 361, Ausgabe B, band 4, VDI Verlag Gmbh, Berlin NW7 (in German).

Reznikov, M.I., Lipov, Yu.M., 1985. Steam Boilers. Mir Publishers, Moscow, 341 p.

VDI heat atlas, 1993. Verein Deutscher Ingenieure, VDI-Gesellschaft Verfahrenstechnik und Chemieingenieurwesen, Düsseldorf, VDI-Verlag, ISBN 3184009157.

5
THERMAL DESIGN OF BOILER PARTS

CHAPTER OUTLINE

Steam Generation from Biomass. DOI: http://dx.doi.org/10.1016/B978-0-12-804389-9.00005-8

The thermal design of boiler parts involves sizing the steam generator components. Sizing includes hydraulic as well as heat transfer sizing. Before the thermal design is undertaken, the general outlook or surface type of the boiler heat transfer surfaces must be known. This means that the order of the heat transfer surfaces with respect to flue gas flow must be ascertained. Usually this information is decided by determining the 'type' of boiler to be designed. Often the furnace is the first heat transfer surface the "flue gas" sees. Heat is first transferred at the superheaters, next at the economizers, and finally at the air preheaters.

In thermal design, first the heat requirement of each surface (temperatures) is set. Then the heat transfer surfaces are sized based on this heat requirement. Sizing includes steam–water side sizing and flue gas side sizing.

Almost always after the first round of heat transfer surface sizing is done, some of the dimensions are changed to arrive at a better solution. The final form and parameters of the design are scrutinized, based on experience of manufacturing, erection, corrosion, past performance, etc., in order to arrive at the most cost-effective solution for a particular customer. Hardly ever can past designs be reused exactly, except in the budgetary phases of a project.

5.1 Furnace Sizing

The main role of the steam generator furnace is to burn the fuel as completely and stably as possible. Leaving unburned material will decrease heat efficiency and increase emissions. Also, the burning must be done in an environmentally sustainable way. The emissions from the furnace must be as low as possible. In addition, the furnace design must be reliable and economical to manufacture. The furnace must be sized to provide as many as possible of the desired features. This is not as simple as it seems as we must balance contradictory requirements.

> *The design of a furnace and the prediction of its performance still owe as much to art, past experience, and empiricism as it does to sound scientific theory.*
>
> **Truelove, 1983.**

It is desirable to arrange combustion equipment in a manner that provides a temperature profile suitable to minimum emission and fouling. The fuel must be burned as completely as possible. The furnace needs to provide the desired exit temperature (over the required operating range). The exit from the furnace must convey combustion products optimally to further heat transfer surfaces.

The main parameters for the furnace sizing are:
- Furnace dimensions
  - height
  - depth
  - width
  - configuration
- Furnace wall construction
- Desired furnace end temperature

There are three engineering methods used to size boiler furnaces:
1. Sizing by scaling and parameters
2. Zero-dimensional models
   a. stirred reactor
   b. profile
3. One-dimensional models

After the furnace has been dimensioned, it is often advantageous to study the furnace operation using a three-dimensional computational fluid dynamics (CFD) model. Typical calculation times for a single furnace range from a day to several weeks. It is impractical to dimension the furnace by CFD alone.

5.1.1 Furnace Dimensions

Most utility and industrial boiler furnaces are rectangular. A large number of package boilers have a cylindrical furnace. The furnace's main dimensions are shown in Fig. 5.1. The main heat transfer mode in the furnace is radiation. Therefore typically the gas velocities in the furnace are lower than on the rest of the surfaces. The desire to increase gas speed can be seen in the design for the furnace nose. The outlet of the furnace is typically 40–70% of the furnace cross-section.

The furnace bottom for typical pulverized boilers is the double-inclined or V-form. A flat bottom ($b \ll w$) is more typical for small oil- and gas-fired units as well as grate, fluidized bed, and recovery boilers. Large oil and gas boilers used to have flat bottoms, but after a series of serious accidents because of high heat fluxes causing steam separation, they are now seldom seen.

In Fig. 5.1 the main dimensions are:

h is the height (typically expressed to the middle of the nose)
b is the height of the furnace's bottom section
w is the width (typically the distance between the tube row centerlines)

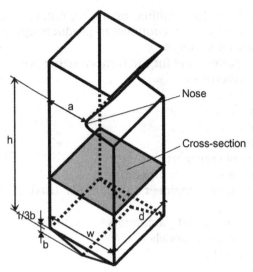

Figure 5.1 Boiler furnace main dimensions.

d is the depth (typically the distance between the tube row centerlines)

a is the nose depth

From these basic dimensions we create more complex definitions.

$$Cross\ section = w \times d \qquad (5.1)$$

$$Exit\ area = a \times d \qquad (5.2)$$

The most typical definition of the available heat transfer surface in the furnace is the effective projected radiant surface (EPRS). By EPRS we mean the surface that is formed by the tube row centerlines.

$$EPRS \sim 2 \times h \times w + 2 \times h \times d \qquad (5.3)$$

Radiant beam length is the effective radiation length for the furnace enclosure. One can think of the radiant beam length as being the average of all possible distances between the surface elements in a boiler furnace.

$$s = \frac{3.6 \times h \times d \times w}{2 \times h \times w + 2 \times h \times d + 2 \times d \times w} \qquad (5.4)$$

where

$$s\ (\text{Radiation beam length}) = 3.6 \times \text{Volume/Area}$$

5.1.2 Typical Furnace Loadings

One of the main methods of evaluating furnace behavior is comparing it to specific reference values. Typical reference furnace loadings can be seen in Table 5.1. The most usual specific reference values are:

1. Effective projected radiant heat flux (EPRH) or heat absorbed by the furnace walls divided by the effective projected surface
2. Heat load per cross-section (heart heat release rate or HHRR) or heat input to the furnace divided by average cross-section
3. Heat load per volume or volumetric heat release rate (VHRR) or heat input to the furnace divided by furnace volume
4. Residence time, typically expressed as furnace volume divided by furnace exit gas flow at the furnace's average temperature and pressure.

It should be noted that the aim of comparing furnace loadings is just that: roughly comparing one furnace to another. Therefore the calculation methods usually do not involve corrections and are coarse.

Effective projected radiant heat can be calculated from

$$EPRH = \frac{\Phi_F}{2 \times h \times w + 2 \times h \times d} \qquad (5.5)$$

where

Φ_F is the heat absorbed by the furnace walls.

Table 5.1 Typical Reference Furnace Loadings

Type	Effective Projected Radiant Heat (EPRH) kW/m^2	Heart Heat Release Rate (HHRR) MW/m^2	Residence Time s
Natural gas	550–630		
Oil	550–630		
Coal, pulverized	220–380		2–2.5
Coal grate	250–410	2–2.2	
Biomass grate		1.2–1.5	
Circulating fluidized bed (CFB)		5	4
Bubbling fluidized bed (BFB)		1.5	8
Recovery	100–280	2.7–3.0	4.5

Heat load per cross-section (the HHRR) can be calculated from

$$HHRR = \frac{\Phi_{net}}{w \times d} \tag{5.6}$$

where

Φ_{net} is the net heat input ($\sim$gross—ash losses and combustion losses)

Heat load per volume (the VHRR) can be calculated from

$$VHRR = \frac{\Phi_{net}}{w \times d \times h} \tag{5.7}$$

It can be seen that oil and gas furnaces have higher wall heat fluxes. Larger units tend to have higher heat release per unit cross-area but lower heat release per unit volume. If we keep the furnace geometry similar and change the furnace dimensions, the relationship between area and volume is

$$A = V^{2/3} \tag{5.8}$$

If we look into specific heat releases per unit volume (Reznikov and Lipov, 1985), we notice that industrial furnaces are large, Table 5.2. This is mainly because their size is limited by the desired heat transfer rate to the walls. It can be concluded that for reasonable combustion, much smaller furnaces could be constructed.

Table 5.2 Heat Release Rates for Different Heat Engines

Type	Main Dimension m	Heat Release kW/m³
Biomass firing	10	100
Cyclone furnace	1–5	1000
Combustion chamber of jet engine	0.2	10,000
Combustion chamber of car engine	0.1	100,000

Source: After Reznikov, M.I., Lipov, Yu.M., 1985. Steam Boilers. Mir Publishers, Moscow, 341 p.

5.1.3 Choice of the Furnace Outlet Temperature

The following factors affect the choice of the furnace outlet temperature:

- Ash characteristics; control of ash behavior at superheaters is a key design parameter
  - first ash melting temperature
  - ash softening temperature
  - ash flowing temperature
- Fuel
  - heating value (low heating value equals low specific loading)
  - specific combustion time (large time means low specific loading)
- Choice of superheater and furnace material
- Desired superheating temperature
- Maximum allowable wall flux taking into account the need of avoidance of departure from nucleate boiling (DNB)
- Effect on emissions (SOx, NOx) (low NOx might require low specific loading)
- Firing method
  - grate
  - pulverized
  - fluidized bed
  - droplet
  - gas

Furnace outlet temperature affects the dimensioning of the rest of the heat transfer surfaces and has a major effect on emissions. It is often practical to find the proper value through several iterative rounds. Typical furnace outlet temperatures can be seen in Table 5.3.

Table 5.3 Typical Furnace Outlet Temperatures at Maximum Continuous Rating Load

Type	Furnace Outlet Temperature (°C)
Coal, high volatiles	950–1000
Peat, pulverized firing	950–1000
Biomass, fluidized bed	1050–1150
Biomass, circulating fluidized bed	900–1000
Peat, pulverized firing	950–1000
Natural gas and oil	900–1200
Recovery boiler	900–1050

5.1.4 Choice of Furnace Wall Material

The most common furnace material is finned carbon steel tube, which forms the membrane wall. Practically all biomass boilers use a carbon steel membrane tube wall construction. For better corrosion resistance a stainless steel or stainless steel outer layer membrane wall tube construction can be used. The stainless steel outer layer is formed by a weld overlay or by using compound tubing. This kind of tube is used in recovery and waste-fired boilers. Typically this outer layer is not considered as a tube for pressure part calculations when required tube thicknesses are determined.

Parts of the furnace wall can be covered by refractory material. Refractory material is used especially in fluidized bed boiler lower furnaces as erosion protection. In grate firing it is used as an insulator to decrease heat transfer, which increases local gas temperature and facilitates combustion reactions. With refuse-fired boilers the refractory material is often used as corrosion protection.

5.1.5 Furnace Air Levels

The choice of furnace air levels and especially the ratio of air to fuel at different furnace locations will affect the burning and so the temperature profile in the furnace. It is typical to introduce some air late, high in the furnace to lower NOx emissions. It is typical for the main combustion to take place close to the zone where fuel is introduced into the furnace. This is typically the hottest zone. It is usual to introduce some air below the main combustion zone though the grate or nozzles (Fig. 5.2).

5.1.6 Circulating Fluidized Bed Furnace Design

When we dimension circulating fluidized bed (CFB) furnaces, it has to be remembered that they contain high amounts of sand, Fig. 5.3. This means that the temperature profile and thus the heat transfer close to the furnace wall differs from that in other types of furnaces.

Furnace gases in the CFB move faster than the sand and fuel particles. The dense matter travels along the walls in a downwards direction. This means that the density profile and heat transfer profile are created. The typical overall heat transfer coefficient is $k = 150-200$ W/m^{2}K.

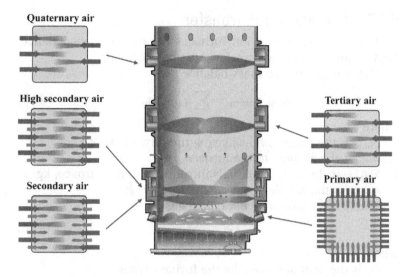

Figure 5.2 Furnace of a large recovery boiler. From Mäntyniemi, J., Haaga, K., 2001. Operating experience of XL-sized recovery boilers. Proceedings of 2001 TAPPI Engineering, Finishing & Converting Conference, TAPPI Press, Atlanta, GA, 7 p. (Mäntyniemi and Haaga, 2001).

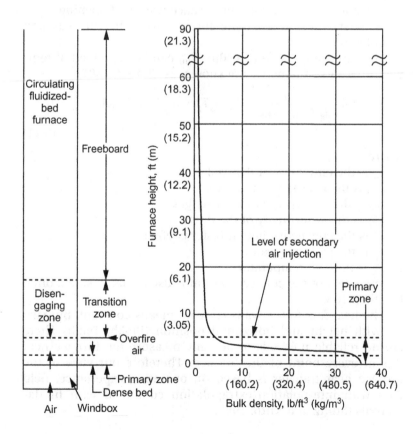

Figure 5.3 Density profiles in a circulating fluidized bed furnace. From Kavidass, S., Alexander, K., 1995. Design considerations of B&W internal circulation CFB boilers. Power-Gen Americas '95, December 5–7, 1995, Anaheim, California, 11 p. (Kavidass and Alexander, 1995).

5.1.7 Furnace Heat Transfer

To be able to dimension furnaces the overall mass balance, heat balance, and heat transfer must be specified.

The overall furnace mass balance is

$$q_{m,fg} = q_{m,a} + \sum q_{m,i} - q_{m,ash} \qquad (5.9)$$

where

$q_{m,fg}$ is the flue gas flow leaving the furnace, kg/s
$q_{m,a}$ is the air flow to the furnace, kg/s
$\Sigma q_{m,f,i}$ is the sum of all fuels introduced to the furnace, kg/s
$q_{m,ash}$ is the ash flow leaving the furnace, kg/s
The furnace gas side heat balance is

$$\Phi_F = \Phi_{net} - \Phi_l - \Phi_e \qquad (5.10)$$

where

Φ_F is the heat absorbed by the furnace walls
Φ_{net} is the net heat input (gross—ash losses and combustion losses)
Φ_l is the heat losses through furnace walls and openings
Φ_e is the heat leaving the furnace with the flue gas and radiation.

The furnace gas side heat flux Φ_{fur} can be expressed, if temperatures and emissivities are known, as (Brandt, 1985)

$$\Phi_F = A_{eff} \frac{\varepsilon_w}{\alpha_{dg} + \varepsilon_w - \alpha_{dg}\varepsilon_w} \delta \left(\varepsilon_{dg} T_g^4 - \alpha_{dg} T_w^4 \right) + \alpha_c A_{eff}(T_g - T_w)$$

$$(5.11)$$

where

ε_w is the emissivity of the wall
A_{eff} is the heat transfer surface
ε_{dg} is the emissivity of the dusty gas
α_{dg} is the absorptivity of the dusty gas
α_c is the convective heat transfer coefficient
T_g is the gas temperature
T_w is the wall temperature.

The effect of the convective term is usually fairly small, often less than 10%.

The heat flux to the walls is by no means constant, but varies with height and horizontal position (Blokh, 1988). There are no reliable methods to actually measure the furnace heat flux at every point of the furnace. Therefore our view of the furnace heat transfer is based on operating experience, field data with interpolation/extrapolation combined with fundamentals (Stultz and Kitto, 1992).

5.1.8 Furnace Model Types

If we observe a bubbling fluidized bed (BFB) furnace or grate-fired boiler furnace, it is easy to see that there are several different zones. Each zone can be characterized by the process that is taking place in it. These models can be described as semiempirical. They combine data from known physical descriptions with experiments. These kinds of model can give a fairly good idea about the effect of some changes, but not all.

Processes in the furnace can be solved by several types of model (VDI heat atlas, 1993). The list of frequently used models includes simple ones as well as the more advanced:

- Zero-dimensional furnace model, stirred reactor model, engineer's model, or black box model. This is the most simple furnace model. The whole process is described by dimensions, input, and output streams.
- One-dimensional furnace model or plug flow model. The furnace is modeled as a plug flow in a tube. This kind of model can take into account the temperature, and species concentration variations in one dimension.
- One-plus-dimensional furnace model. Especially for CFB calculations the plug flow furnace models have been improved. One can include the return flow loop. Often gas and inert material flows are separated. This kind of model can take into account the temperature, and species concentration variations in one dimension.
- Three-dimensional computational fluid model (the CFD). This is the most detailed model. The furnace is divided into hundreds of thousands of little volumes. Heat, mass, and momentum balances are solved for each volume. These balances can be coupled with chemical reactions and heat transfer in different ways.

5.1.9 Furnace Dimensioning, Stirred Reactor

One of the most used furnace dimensioning methods is the stirred reactor model. In it the furnace is approximated to be filled with homogenous gray gas at uniform temperature and pressure. Often the gray gas is approximated with a three-gas and dust mixture. All furnace phenomena, especially heat transfer, are calculated assuming uniform temperature. At the furnace exit the temperature is decreased by a specified amount.

The stirred reactor furnace dimensioning can be done as follows:

1. Guess the initial furnace dimensions: shape, height, width, depth.

Table 5.4 Stirred Reactor, ΔT to Use

Boiler Type	ΔT °C
Pulverized coal-fired	100–200–300
Grate-fired	100–<u>130</u>–180
Oil- and gas-fired	100–150–200
Bubbling fluidized bed	100–<u>130</u>–150
Circulating fluidized bed	0
Recovery	N.A.

2. Guess the furnace exit temperature T_e.
3. Calculate the heat transfer using $T_{fg} = T_e + \Delta T$.
4. Calculate the furnace exit temperature from the heat balance with the calculated heat transfer.
5. If not converged, then repeat from step 2.
6. If the desired furnace exit temperature is not achieved, repeat from step 1.

The typical values of ΔT to use for the different types of furnace can be seen in Table 5.4. The stirred reactor model is not suitable for dimensioning the recovery boiler furnace.

5.2 Superheater Dimensioning

Superheater dimensioning is based on determining appropriate radiative and convective heat transfer as well as ash deposit properties on the superheater surface. Superheater dimensioning plays a major role in boiler cost as superheaters are typically manufactured from much more expensive alloyed steel tubes than other boiler parts (Nishikawa, 1999). Superheaters are often the most expensive heat transfer surface. Heat transfer in superheaters is mainly radiative, but in primary superheaters, convection often plays a major role.

The superheater arrangement must be built so that it superheats the required amount, from low to high loads. This can be achieved by the correct choice of radiative and convective superheating surfaces. The last superheater cannot have too large a temperature rise. Often the last superheater section transfers 20–33% of the total superheating heat requirement

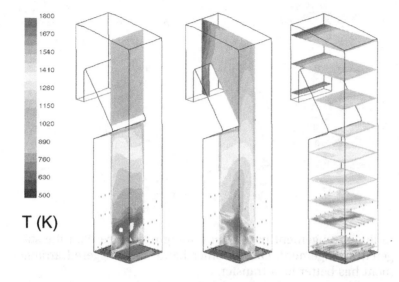

T (K)

Figure 5.4 Temperature profile in a recovery boiler. From Vakkilainen, E.K., 2005. Kraft recovery boilers—principles and practice. Suomen Soodakattilayhdistys r.y., Valopaino Oy, Helsinki, Finland, 246 p. ISBN 9529186037 (Vakkilainen, 2005).

(Steinwall et al., 2002). Superheater design should also be such that temperature differences between adjacent tubes are small. In addition, particular attention should be placed on superheaters that receive furnace radiation.

Temperature differences can be somewhat controlled by changing the tube lengths between passes. The outermost tube, which receives the most radiative flux, should be shorter than the rest of the tubes. A proper superheater arrangement also eliminates many of the problems to do with uneven or biased flue gas flow (Fig. 5.4).

One of the most difficult tasks is to account for uneven temperature and flow fields. An example of CFD calculation can be seen in Fig. 5.7. Typically the superheater needs to be reduced to a series of heat transfer surfaces, which are then solved separately.

5.2.1 Tube Arrangement and Spacing

An in-line tube arrangement, Fig. 5.5 is preferred for boilers where fouling resistance is important (Teir, 2004). Such boilers are grate-fired boilers, solid fuel BFB and CFB boilers like bark boiler and kraft recovery boilers. The staggered arrangement is often only used in oil, gas, and heat recovery boilers. The free space between tubes is much smaller with the staggered arrangement than with the in-line arrangement. Therefore the

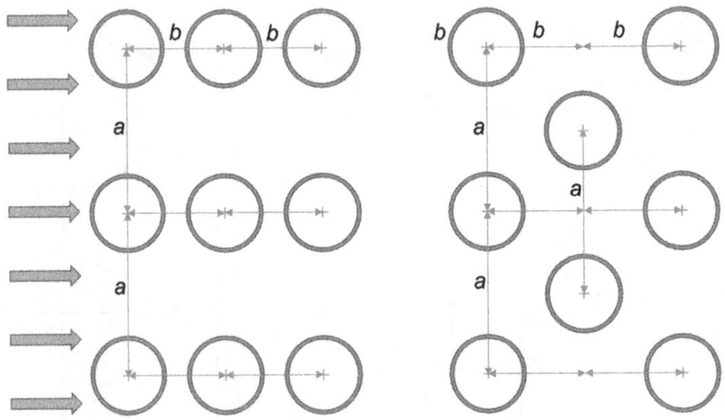

Figure 5.5 Superheater tube arrangement, with flue gas flowing across the tubes from the left with transversal pitch *a* and longitudinal pitch *b*, Left: inline, Right: staggered.

in-line arrangement has better fouling resistance than the staggered arrangement. On the other hand, the staggered arrangement has better heat transfer.

For low temperatures and for a nonfouling zone a tight spacing can be used. For high temperature or where fouling is probable, wide spacing is used.

5.2.2 Panel Superheater

A superheater where tubes are placed in an in-line configuration and close to each other in the flue gas flow direction is called a panel superheater, Fig. 5.6. In panel superheaters the tube longitudinal spacing is smaller than $1.25 \times \emptyset_d$. The panel superheater is more resistant to fouling and erosion and can withstand a high heat flux. It is used in the most demanding applications, typically as the first flue gas side superheater after the furnace.

5.2.3 Backpass Superheater Set

The section of the boiler that is formed when the flue gas is turned to flow downwards after the furnace and the possible superheating section on top of the furnace is called the backpass. Superheaters that are located after the flue gas starts flowing downwards are called backpass superheaters. In large CFB and BFB units it is common to use a horizontal tube arrangement. This means that almost all the tubes are placed in a horizontal position. Backpass superheater tubes hang from the backpass roof.

For large reheater units the backpass is sometimes divided into two sections to improve superheating control.

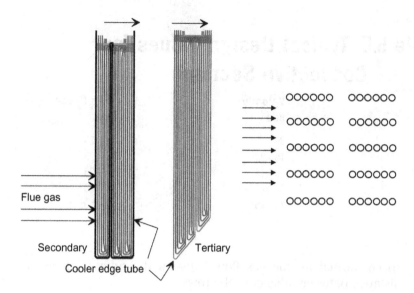

Figure 5.6 Panel superheater.

5.3 Convective Section Dimensioning

In this section the boiler bank, economizer and air heater dimensioning are described. For each application, typical dimensioning parameters are discussed. When the flue gas temperature decreases, the fraction of heat transfer that is attributed to radiation decreases. It is common practice that in surfaces where the average flue gas temperature is at or below 500°C the contribution from radiative heat transfer can be assumed to be much below 10%. When the average flue gas temperature is below 500°C, the heat transfer surfaces are deemed to be placed in the boiler convective section.

5.3.1 Design Velocity

Design velocity and spacing for convective heat transfer surfaces can be seen in Table 5.5. If we place tubes close to each other and use high flue gas velocity, then the propensity to foul and erode is high. Especially with biofuels there is a need to utilize moderate flue gas velocities and have ample spacing. When we look at these values, the point of interest is usually taken as the smallest cross-section between tubes at average flue gas conditions in the middle of the heat transfer surface. This means that in a large heat transfer surface the entry conditions are somewhat worse. Spacing means distance between adjacent tube rows. Often the determining parameter is expressed as free

Table 5.5 Typical Design Values for Convective Section

Fuel Type	Velocity m/s	Free Space mm
Gas, light, and heavy oil	30	50—25
Bituminous low ash coal	15—20	40—25
Bituminous high ash coal	10—15	50—25
Peat, bark	8—12	50—25
Recovery	10—15	70—50

space normal to flue gas flow between the tubes and not as distance between adjacent tube rows.

5.3.2 Boiler Banks

Boiler banks, steam generating banks, or evaporative heat transfer surfaces are typical in low-pressure and small boilers. They are almost always used in heat recovery steam generator (HRSG) boilers. Typically the furnace is dimensioned to capture enough heat from flue gases to evaporate all the steam. In low-pressure boilers this would mean too low a furnace exit temperature and too large superheater surfaces. Therefore some of the evaporative duty is transferred to boiler banks.

Boiler banks transfer heat from flue gas to evaporate the steam—water mixture. A two-drum boiler bank was very typical in early boilers and is still used in smaller, low-pressure boilers. In small biomass BFB boilers one can often find a boiler bank (i.e., an evaporative heat transfer surface). To keep the steam—water mixture flowing, the tubes are often inclined 5—15 degrees from horizontal.

5.3.3 Economizers

Economizers are one of the four main heat transfer surface types. The name economizer comes from German *Ekonomizierung* (to economize). They were first used in shell and tube boilers to preheat the feedwater. Feedwater preheating helps to capture more heat from the flue gases. Economizers are usually convective heat transfer surfaces. They are often made with finned tubes, especially in HRSGs.

In solid biomass boilers and recovery boilers, straight tubes are often preferred as finned tubes are more prone to fouling. Economizers rarely suffer from corrosion or erosion. Therefore tighter spacing than for superheaters can be used.

5.3.4 Air Preheaters

There are two main types of air heaters. Recuperative air heaters are tube heat exchanges where hot flue gases heat combustion air. Recuperative heat transfer surface arrangements are practically the only alternative in biomass boilers below 500 MWth. Often in the recuperative air preheater the flue gas heats combustion air flowing inside the tubes, but also air heaters where flue gas is in the tubes have been used. Because the heat transfer is from gas to gas by convection, the overall heat transfer coefficient is low and the surface requirement is large. Because of this, in large boilers regenerative air preheaters can be used. Because of higher air pressure, regenerative air heaters are seldom used in fluidized bed boilers. The main differences between recuperative and regenerative air heaters are (Lipetz et al., 2002):

- Recuperative air heaters are cheaper in boilers below 500 MWth.
- Recuperative heaters have a good record of use.
- Regenerative air heaters are lighter and require less space.
- There is a high air leakage in regenerative air heaters to flue gas side corresponding from 8% to 15% of air flow, depending on the condition of seals.
- Ash is entrained with air in the regenerative heater.
- Recuperative air heaters often have better heat efficiency.

One of the most typical tubular air heater configurations is the two-pass air heater. Thermal performance of a two-pass recuperative air heater is between cross-flow and parallel flow. The multipass recuperative air heater is also very typical. The four-pass recuperative air heater is thermally comparable to a counterflow arrangement. The air preheater should be designed not to foul easily.

The regenerative air heater uses the heat capacity of inert mass to transfer heat. Flue gas heats ceramic or metallic mass. This mass in turn heats incoming air. Regenerative air preheaters can be used up to about 400°C flue gas temperature. Regenerative air preheaters occupy little space, about a quarter or a sixth of that required by recuperators. They have a small temperature difference and are the only type of air preheater that permits a large heat-exchanging surface to be produced cheaply.

Because there must always be some clearance between the rotor and the casing of the rotating heating surface, leakage cannot be avoided. In operation the leakage is increased by the warping of the rotor. This is a weakness of regenerative air preheaters because it increases the power consumption of draught fans.

5.4 Heat Transfer in Boilers

The purpose of this section is to give the reader an overview of some possible methods to calculate heat transfer in boilers. The treatment is short and the reader is requested to study other material if they are interested in learning about heat transfer phenomena in more detail. Typically the preliminary design of biomass boilers is done using similar methods to those presented here. Utilizing methods with higher accuracy, requiring more detailed input, is often futile as the calculated value must be corrected with a large fuel- and boiler-type specific correction factor because of additional heat transfer resistance by ash at boiler surfaces.

5.4.1 Overall Heat Transfer

Overall heat transfer in a boiler heat transfer surface can be expressed with the general heat transfer equation as

$$\Phi = kA\Delta T \tag{5.12}$$

where

Φ is the heat transferred

k is the overall heat transfer coefficient

A is the heat transfer surface

ΔT is the temperature difference

The overall heat transfer coefficient is a function of convective and radiative heat transfers.

$$k = \frac{f_o}{\dfrac{1}{k_i} + \dfrac{1}{k_o} + F(\lambda, d_o, d_s, s)} \tag{5.13}$$

$$k_o = f_n k_c + k_r + k_{ex}$$

where

f_n is the form correction (correction for number of rows, correction for arrangement, etc.)

f_o is the overall correction (fouling correction, etc.)

k_i is the inside heat transfer coefficient referred to the outside surface

k_o is the outside heat transfer coefficient

k_r is the radiative heat transfer coefficient

k_{ex} is the external heat transfer, arranged to represent a heat transfer coefficient

k_c is the convective heat transfer coefficient

λ is the heat conductivity of tube material

d_o is the outside tube diameter

d_s is the inside tube diameter

s is the tube wall thickness

The correction for the number of rows, f_n, can be expressed as

$$f_n = \left\{ \begin{array}{l} \dfrac{(n_r-1)^2 - 1}{(n_r-1)^2} ; \; n_r > 1 \\[2mm] 0.75 ; \; n_r = 1 \end{array} \right\} \tag{5.14}$$

The correction for overall fouling, temperature deviations, flow deviations, etc., f_o, is usually dependent on fuel type and properties, steam generator configuration, type of surface, and expected calculation error. The exact numerical value must therefore be determined empirically from previous experimental data.

The heat transfer resistance through a tube of uniform material for a tubular construction where tubes are separated from each other is

$$F(\lambda, d_o, d_s, s) = \frac{1}{\frac{d_o}{2\lambda} \ln \frac{d_o}{d_i}} \tag{5.15}$$

5.4.2 Radiation Heat Transfer

Radiation heat transfer can be calculated according to various heat transfer functions. It has been typical to fit experimental data to simple or to more complex equations, which can be found in heat transfer books. The formulas used here can be found in *VDI—Wärmeatlas*, Kc1–Kc12. Even though the concept of the radiation heat transfer coefficient has only a very weak connection to physical reality, it is used here so that the radiative heat transfer can conveniently be expressed with the convective heat transfer using Eq. 5.12. The radiation heat transfer coefficient can be determined through radiation heat flow.

$$k_r = \frac{\Phi_r}{A_{eff}(T_g - T_w)} \tag{5.16}$$

The radiation heat flow Φ_r can be expressed if temperatures and emissivities are known as

$$\Phi_r = A_{\text{eff}} \frac{\varepsilon_w}{\alpha_{dg} + \varepsilon_w - \alpha_{dg}\varepsilon_w} \delta(\varepsilon_{dg} T_g^4 - \alpha_{dg} T_w^4) \tag{5.17}$$

where

ε_w is the emissivity of the wall; for fully oxidized boiler tubes 0.8 is usually used

ε_{dg} is the emissivity of the dusty gas

α_{dg} is the absorptivity of the dusty gas

For dusty gas we can express the emissivity and absorptivity, assuming no band overlapping, through pure gas and pure dust emissivities

$$\varepsilon_{dg} = \varepsilon_g + \varepsilon_d - \varepsilon_g \varepsilon_d \tag{5.18}$$

$$\alpha_{dg} = \alpha_w + \varepsilon_d - \alpha_w \varepsilon_d \tag{5.19}$$

The emissivities and absorptivities of pure gas mixtures can be calculated with pure gas properties

$$\varepsilon_g = \varepsilon_{H_2O}(T_g, sp_{H_2O}) + \varepsilon_{CO_2}(T_g, sp_{CO_2}) \\ + \varepsilon_{SO_2}(T_g, sp_{SO_2}) + \Delta\varepsilon_g(\varepsilon_{H_2O}, \varepsilon_{CO_2}, \varepsilon_{SO_2}) \tag{5.20}$$

$$\alpha_{dg} = \alpha_{H_2O} + \alpha_{CO_2} + \alpha_{SO_2} - \Delta\alpha_g \tag{5.21}$$

where

$\Delta\varepsilon_g$ is the overlapping correction for emissivity

$\Delta\alpha_g$ is the overlapping correction for absorptivity

p_x is the partial pressure of substance x

s is the radiation beam length

As no better approximations have been developed, we must express absorptivities through emissivity functions (Coelho, 1999), applicable when total gas side pressure is about 0.1 MPa.

$$\varepsilon_g = (0.595 - 0.00015 \times T_G)(1 - e^{-0.824sp_{tot}}) \\ + (0.275 - 0.000115 \times T_G)(1 - e^{-25.907sp_{tot}}) \tag{5.22}$$

$$p_{tot} = p_{H_2O} + p_{CO_2}$$

$$\alpha_{H_2O} = \left(\frac{T_g}{T_w}\right)^{0.45} \varepsilon_{H_2O}\left(T_g, sp_{H_2O} \frac{T_w}{T_g}\right) \tag{5.23}$$

$$\alpha_{CO_2} = \left(\frac{T_g}{T_w}\right)^{0.65} \varepsilon_{CO_2}\left(T_g, sp_{CO_2} \frac{T_w}{T_g}\right) \tag{5.24}$$

$$\alpha_{SO_2} = \left(\frac{T_g}{T_w}\right)^{0.50} \varepsilon_{SO_2}\left(T_g, sp_{SO_2}\left(\frac{T_w}{T_g}\right)^{1.5}\right) \tag{5.25}$$

The individual emissivity functions are shown in Appendix A and can be mathematically approximated as

$$\varepsilon_{CO_2} = \alpha_0 + \alpha_1\mu + \alpha_2\mu^2 + \alpha_3\mu^3 \tag{5.26}$$

$$\mu = \mu(T_g) \tag{5.27}$$

$$\varepsilon_{H_2O} = \varepsilon_\alpha\left(1 - \varepsilon^{(f(ps)g(ps,T_g))}\right) \tag{5.28}$$

$$\alpha_i = \alpha_i(p_{CO_2}, s) \tag{5.29}$$

$$\varepsilon_{SO_2} = \alpha\left(1 - e^{kT-b}\right)\left(sp_{SO_2}\right)^{0.87} \tag{5.30}$$

where ε_α is the background emissivity.

For dust emissivity ε_d a fully dispersed approach is used

$$\varepsilon_d = 1 - \varepsilon^{-\alpha_d B_d s} \tag{5.31}$$

where

α_d is the emission coefficient of dust
B_d is the dust loading, kg/m^3
s is the radiation beam length

The expression for radiation beam length s is very complicated and involves the integrating in space of the surface element. A table is provided in Appendix B for radiation beam length.

5.4.3 Outside Convection Heat Transfer

Convection heat transfer calculation is typically based on the expansion of single tube row heat transfer to multiple rows. Single tube row heat transfer is often approximated by various heat transfer equations (Brandt, 1985). Note that, for wider applicability, the laminar flow region equation and the turbulent flow region equation are bound together in a single equation.

$$Nu = f_O f_A Nu_1 = f_O f_A(0.3 + \sqrt{Nu_{lam}^2 + Nu_{tur}b^2}) \tag{5.32}$$

$$Nu_{lam} = 0.664\sqrt{Re}\sqrt[3]{Pr} \tag{5.33}$$

$$Nu_{\text{turb}} = 0.037 \frac{Re^{0.8} Pr}{1 + 2.443 Re^{-0.1}(Pr^{2/3} - 1)} \tag{5.34}$$

The correction factor f_O for outside heat transfer is an empirical coefficient. It is dependent on fuel type and properties, steam generator configuration, type of surface, used convection heat transfer correlation, and expected calculation error. The exact numerical value must be determined empirically from previous experimental data on full-scale boilers. However, we can determine the portion f_A that depends on dimensionless tube arrangement by in-line tubes as

$$f_A = 1 + \frac{0.7 \left(\dfrac{b}{d} \Big/ \dfrac{a}{d} - 0.3\right)}{1 - \dfrac{\pi a}{4 d} \left(\dfrac{b}{d} \Big/ \dfrac{a}{d} + 0.7\right)^2} \tag{5.35}$$

And for staggered arrangement as

$$f_A = 1 + \frac{2}{3b/d} \tag{5.36}$$

See Fig. 5.5 for a and b.

5.4.4   Gas Side Pressure Drop, In-line

Gas side pressure drop can be calculated according to the *VDI Wärmeatlas*, blatt Ld.

$$\Delta p_{\text{gs}} = \Delta p_{\text{f}} \tag{5.37}$$

Pressure losses = friction losses

For an in-line arrangement the pressure drop coefficient for the heat transfer surface with horizontal tubes is

$$\Delta p = n_r \zeta_r \Delta p_d \tag{5.38}$$

where
   n_r is the number of rows in the heat transfer surface
   Δp_d is the dynamic pressure, calculated at the gas side mean temperature and smallest area.
   The single row pressure drop ζ_r for in-line tubes is

$$\zeta_r = \zeta_l + \zeta_t \left(1 - e^{-\frac{Re-1000}{2000}}\right) \tag{5.39}$$

$$\zeta_l = \frac{280\pi((s_l^{0.5} - 0.6)^2 + 0.75)}{(4 s_t s_l - \pi) s_t^{1.6} Re} \tag{5.40}$$

$$\zeta_t = 10^{0.47(s_t/s_l - 1.5)} \left[(0.22 + 1.2) \frac{\left(1 - \frac{0.94}{s_l}\right)^{0.6}}{(s_t - 0.85)^{1.3}} + 0.03(s_t - 1)(s_l - 1) \right]$$

$$(5.41)$$

where

ζ_l is the laminar part of the pressure drop coefficient
ζ_t is the turbulent part of the pressure drop coefficient
s_t is the dimensionless transverse pitch, $s_t = a/d_o$
s_l is the dimensionless longitudinal pitch, $s_l = b/d_o$
Re is the Reynolds number, calculated at the gas side mean
temperature and smallest area
Gas side pressure drop, staggered
The single row pressure drop ζ_r for staggered tubes is

$$\zeta_r = \zeta_l + \zeta_t \left(1 - e^{-\frac{Re-200}{1000}} \right) \qquad (5.42)$$

$$\zeta_l = \frac{280\pi \left(\left(s_l^{0.5} - 0.6 \right)^2 + 0.75 \right)}{(4s_t s_l - \pi)c^{1.6} \, Re} \qquad (5.43)$$

$$c = s_t \ ; \ s_l \geq \sqrt{2s_t - 1}/2$$

$$c = \sqrt{\left(\frac{s_t}{2} \right)^2 + s_l^2} \ ; \ s_l < \sqrt{2s_t - 1}/2 \qquad (5.44)$$

$$\zeta_t = 2.5 + \left(\frac{1.2}{(s_t - 0.85)^{1.08}} \right) + 0.4 \left(\frac{s_l}{s_t} - 1 \right)^3 - 0.01 \left(\frac{s_t}{s_l} - 1 \right)^3$$

5.4.5   Inside Heat Transfer, Tube Fluid

Turbulent convection heat transfer for steam and water flowing inside a circular tube can be calculated according to
the Hausen equation (Hausen, 1974) ; $2300 < Re < 10^7, 0.6 < Pr$
$< 500, d \ll L, T_w \approx T_f$

$$k_i = \frac{\lambda Nu}{d_o} \qquad (5.45)$$

$$Nu_i = 0.0235(Re^{0.8} - 230)(1.8Pr^{0.3} - 0.8) \qquad (5.46)$$

For a laminar flow case, Re < 2300, a value of

$$Nu_i = 3.64 \qquad (5.47)$$

can be used. Usually no correction for inside fouling is used, as the tube inside is kept relatively clean during normal boiler operation.

5.5 Example Calculation of Heat Transfer Surface

We have a single heat transfer surface arrangement, Fig. 5.7 consisting of the main heat transfer surface and side walls situated transverse to the gas flow.

We can make the following simplifications applicable to most of the cases in practice:

- dust is dispersed uniformly to the gas flow
- temperatures of the dust T_{di} and T_{do} are equal to the respective temperatures of the gas flow T_{gi} and T_{go}
- the temperature of the side walls T_{sw} can be treated as constant, i.e., $T_{gi} - T_{go} \gg T_{swo} - T_{swi}$
- no combustion inside the heat transfer surface volume
- heat losses Φ_l and external heat flow Φ_{ex} can be determined separately from the superheater heat transfer calculations

Figure 5.7 Boiler superheater arrangement.

We want to determine the heat transfer from the flue gas flow to the steam flow. The basic equation involved is

$$\Phi_s = \int k(T_g - T_s)dA \qquad (5.48)$$

In the case of superheaters we usually determine an overall heat transfer coefficient k_t, which is treated as an independent value. The temperature distribution as a reference to area can be solved for simple parallel and counterflow cases. Usually we express the solved equation in a basic form.

$$\Phi_s = \int k_t \theta_{\ln} A_{eff} \qquad (5.49)$$

Heat to steam flow = overall heat transfer coefficient × logarithmic temperature difference × effective heat transfer surface, where

$$\theta_{\ln} = \theta_{\ln}(T_{gi}, T_{go}, T_{si}, T_{so}) \qquad (5.50)$$

To aid us in heat transfer calculations, when solving unknown temperatures for a known area and overall heat transfer coefficient, we use nondimensional heat transfer parameters ε, R, and z (Ryti, 1969).

$$\varepsilon = \varepsilon(T_{gi}, \ T_{go}, \ T_{si}, \ T_{so})$$
$$R = R(T_{gi}, \ T_{go}, \ T_{si}, \ T_{so}) \qquad (5.51)$$
$$z = z(T_{gi}, \ T_{go}, \ T_{si}, \ T_{so})$$

We do this by starting from the definition of the heat flow to the steam.

$$\Phi_s = q_{ms} c_{ps}(T_{so} - T_{si}) \qquad (5.52)$$

For the simple no external heat flow case the heat flow to the steam equals the heat flow from the flue gas

$$\Phi_g = q_{mg} c_{pg}(T_{gi} - T_{go}) = \Phi_s \qquad (5.53)$$

Then the parameters describing the heat transfer can be defined

$$G = k_t A_{eff} \qquad (5.54)$$

$$C_{max} = Max(q_{mg} c_{pg}, q_{ms} c_{ps})$$

$$C_{min} = Min(q_{mg} c_{pg}, q_{ms} c_{ps})$$

$$\Delta T_{max} = Max(T_{gi} - T_{go}, T_{so} - T_{si})$$

$$\Delta T_{min} = Min(T_{gi} - T_{go}, T_{so} - T_{si})$$

$$\theta = Max(T_{gi}, T_{go}, T_{so}, T_{si}) - Min(T_{gi}, T_{go}, T_{so}, T_{si})$$

$$\varepsilon = \frac{\Delta T_{max}}{\theta}$$

$$R = \frac{\Delta T_{min}}{\Delta T_{max}}$$

$$z = \frac{G}{C_{min}}$$

$$G = k_t A_{eff}$$

$$C_{max} = Max(q_{mg} c_{pg}, q_{ms} c_{ps})$$

$$\theta = Max(T_{gi}, T_{go}, T_{so}, T_{si}) - Min(T_{gi}, T_{go}, T_{so}, T_{si})$$

$$\varepsilon = \frac{\Delta T_{max}}{\theta}$$

$$R = \frac{\Delta T_{min}}{\Delta T_{max}}$$

$$z = \frac{G}{C_{min}}$$

Then we can express the heat flow to steam also with the dimensionless parameters

$$\Phi_s = \Phi_s(\varepsilon, R, z) = q_{ms} c_{ps}(T_{so} - T_{si}) \qquad (5.55)$$

For the complex heat transfer situation occurring in the recovery boiler superheaters, where there are also several additional heat flows, we can redefine the nondimensional heat transfer parameters with the aid of correction factor b to be able to treat this multiheat flow case as a single heat flow case in the heat transfer calculations

$$G = k_t A_{eff} \qquad (5.56)$$

$$\theta = T_{gi} - T_{si}$$

$$R = \frac{b(T_{so} - T_{si})}{T_{gi} - T_{go}}$$

$$\theta_{ln} = \frac{(T_{gi} - T_{si}) - (T_{go} - T_{so})}{\ln\dfrac{T_{gi} - T_{si}}{T_{go} - T_{so}}}; \text{ parallel flow}$$

$$\theta_{ln} = \frac{(T_{gi} - T_{so}) - (T_{go} - T_{si})}{\ln\dfrac{T_{gi} - T_{so}}{T_{go} - T_{si}}}; \text{ countercurrent flow}$$

$$\varepsilon = 1 - \frac{1 - e^{-z(1+R)}}{1 + R}; \text{ parallel flow}$$

$$\varepsilon = 1 - \frac{1 - R}{e^{z(1+R)} - R}; \text{ countercurrent flow}$$

$$\theta_{sw} = \frac{T_{gi} - T_{go}}{\ln\dfrac{T_{gi} - T_{sw}}{T_{go} - T_{sw}}}$$

$$z = \frac{G}{c_{pg}q_{mg}}$$

It must be noted that the above is valid only for the case when flue gas heats the steam.

5.5.1 Heat Balance

For the superheater the first principle must hold true

$$\Phi_g - \Phi_s - \Phi_{sw} + \Phi_{ex} - \Phi_l = 0 \tag{5.57}$$

Heat from gas − heat to steam − heat to side walls + external heat − heat losses = 0

where

$$\Phi_s = G\theta_{ln} = q_{ms}c_{ps}(T_{so} - T_{si})$$
$$\Phi_g = q_{mg}c_{pg}(T_{gi} - T_{go}) + q_{md}c_{pd}(T_{di} - T_{do})$$
$$= (q_{mg}c_{pg} + q_{md}c_{pd})(T_{gi} - T_{go}) \tag{5.58}$$
$$\Phi_{sw} = k_{sw}A_{sw}\theta_{sw}$$

External heat Φ_{ex} and heat losses Φ_l are usually given and so remain fixed.

5.5.2 Heat Capacities

To make use of the dimensionless parameters ε, R, and z possible we must refer the heat transfer situation to the simple two massflows exchange heat case with the aid of correction factor b. We have chosen the steam side heat flow Φ_s to be the determining heat flow. Then for superheaters

$$C_{min} = q_{mg}c_{pg}$$
$$C_{max} = q_{ms}c_{ps}$$

(5.59)

$$C_{min}^* = bq_{mg}c_{pg}$$
$$R = \frac{C_{min}^*}{C_{max}}$$

$$b = \frac{c_{pg}q_{mg} + \dfrac{\Phi_{ex} - \Phi_l - \Phi_{sw}}{T_{gi} - T_{go}} + c_{pd}q_{md}}{c_{pg}q_{mg}}$$

In practice the coefficient b does not change much as a function of heat transferred, so this type of equation allows one to quickly iterate heat transfer in the case of multiple surfaces.

References

Blokh, A.G., 1988. Heat Transfer in Steam Boiler Furnaces. Hemisphere Publishing Corporation, p. 283. ISBN 0891166262.

Brandt, F., 1985. Wärmeübertragung in dampferzeugern und wärmeaustauschern (Heat transfer in steam boilers and heat exchangers). Vulkan-Verlag, Essen, 281 p. ISBN 3802722744 (in German).

Brandt, F., 1999. Dampferzeurger, kesselsysteme energiebilanz stömungstechnik. (Steam generators, boiler systems, energy balances, fluid mechanics). Vulkan-Verlag, Essen, 283 p. ISBN 3802735048 (in German).

Coelho, P.J., 1999. An engineering model for the calculation of radiative heat transfer in the convection chamber of a utility boiler. J. Inst. Energy. 72, 117–126.

Hausen, H., 1974. Erweiterte gleichung für den wärmeübergang bei turbulenter strömung. (Expanded equation for heat transfer in turbulent flow). Wärme- und Stoffübertragung. 7, 222–225 (in German).

Kavidass, S., Alexander, K., 1995. Design considerations of B&W internal circulation CFB boilers. Power-Gen Americas '95, December 5–7, 1995, Anaheim, California, USA, 11 p.

Lipetz, A.U., Kuznetsova, S.M., Dirina, L.V., 2002. Rotary regenerative or recuperative tubular air heaters. VGB Power Tech. 82 (7), 53–60.

Mäntyniemi, J., Haaga, K., 2001. Operating experience of XL-sized recovery boilers. Proceedings of 2001 TAPPI Engineering, Finishing and Converting Conference, TAPPI Press, Atlanta, GA, 7 p.

Nishikawa, E., 1999. General planning of boiler gas side heat transfer surfaces. In: Ishigai, S. (Ed.), Steam Power Engineering. Cambridge University Press, p. 394. ISBN 0521626358.

Reznikov, M.I., Lipov, Yu.M., 1985. Steam Boilers. Mir Publishers, Moscow, p. 341.

Ryti, H., 1969. Lämmönsiirto (Heat transfer). Tekniikan käsikirja, Osa 4, K. J. Gummerus, Jyväskylä, pp. 595–600 (in Finnish).

Steinwall, P., Johansson, K., Svensson, S.-Å., Adolfson, H., Ekelund, N., 2002. Optimum steam data in bio-fuelled CHP. (Optimala ångdata för biobrensleeldade kraftwärmeverk). Report Number SVF-770, Project Värmeforsk-A9-830, 61 p (in Swedish).

Stultz, S.C., Kitto, J.B., (eds.), 1992. Steam Its Generation and Use, 40th ed. 929 p. ISBN 0963457004.

Teir, S., 2004. Steam Boiler Technology, second ed. Energy Engineering and Environmental Protection publications, Helsinki University of Technology, Department of Mechanical Engineering, p. 215. ISBN 9512267594.

Truelove, J.S., 1983. Furnaces and combustion chambers. In: Heat Exchanger Design Handbook, HEDH, Ed. Ernst U. Schlünder Part 3. Thermal and hydraulic design of heat exchangers. VDI, Düsseldorf. ISBN 3184190838.

Vakkilainen, E.K., 2005. Kraft recovery boilers—principles and practice. Suomen Soodakattilayhdistys r.y., Valopaino Oy, Helsinki, Finland, 246 p. ISBN 9529186037.

VDI heat atlas, 1993. Verein Deutscher Ingenieure, VDI-Gesellschaft Verfahrenstechnik und Chemieingenieurwesen. Düsseldorf, VDI-Verlag, ISBN 3184009157.

6

AUXILIARY EQUIPMENT

In addition to a building, pressure part, and electrical equipment the steam boiler plant consists of a number of items of miscellaneous machinery. These are called auxiliary equipment because they help the operation of the main equipment.

Steam Generation from Biomass. DOI: http://dx.doi.org/10.1016/B978-0-12-804389-9.00006-X

6.1   An Overview of Auxiliary Equipment

Normally a steam boiler plant includes additional machinery to enable the operation. Such equipment is typically bought from third parties and includes (Akturk et al., 1991):

- burners
- fans, ducts, dampers
- air heaters
- hoppers, silos, crushers
- sootblowers, conveyors

To be able to design and operate a steam boiler one needs to understand the functioning of different pieces of equipment. It is recognized that for steam boiler design it is more important to understand the purpose and limitations of each piece of equipment than to be able to perform a detailed design of this equipment (Basu and Fraser, 1991) . Therefore you will find that pumps, blowers, fans, and various flue gas cleaning devices, e.g., are discussed more thoroughly in other relevant literature.

Solid fuels typically fired in industrial and utility boilers include: bituminous coal, lignite (brown coal), wood, paper sludge, bagasse, bark, peat, and refuse-derived fuel (RDF). Each fuel needs some consideration of its own peculiarities, but generally all can be fired in multiple ways even though this does not always make economical or technical sense. The most typical solid fuel firing methods are grate firing, pulverized firing, and fluidized bed firing. All solid fuel firing systems need fuel handling and storage devices (Brindle et al., 1991). In addition, proper ash handling equipment is necessary.

6.1.1   Auxiliary Equipment for Pulverized Firing

The pulverized firing of solid fuels is mostly done with coal. In the past, coal was burned as lumps (Croft, 1922), but now almost all coal for electricity generating is burned as about 0.1 mm particles (Stultz and Kitto, 1992). Pulverized firing can be used with a very large unit size, up to 1000 MW_e. The main disadvantage in pulverized firing is that it is very hard to reduce emissions during combustion, so additional units for SOx and NOx control are required (Doležal, 1985). The main special requirement for pulverized firing is that the fuel particle size is small so the burning time is fast and reaction rate is increased due to the higher surface area when the particle size is reduced.

If the biomass is already of small size (<1 mm), e.g., sawdust or byproducts from alimentary industries, then pulverized firing is economical. In most cases it takes too much electrical energy

(30–300 kWh/tonne) to decrease the biomass particle size down to what is required. In some special cases the electricity consumption of crushing the wood can be economical (Wadsborn et al., 2007). Pellets and small-sized biomass have been cofired in pulverized coal boilers (Livingston, 2013). Due to the particle size, the unburned amount tends to be too high after the fraction of the biomass reaches 5–10% of the boiler's required heat input. If the biomass is hard and fragile, such as torrefied biomass, the ratio can be significantly increased (Bergman et al., 2005). It seems that the fraction of the biomass can be 30–50% of the boiler's required heat input.

Typical pieces of auxiliary equipment needed for the pulverized firing of solid fuels are ash handling, special burners, biomass crushers (hammermills), biomass handling, fans, blowers, and pumps. For the pulverized firing of solid biomass one typically also has emission control equipment, such as an electrostatic precipitator.

6.1.2 Auxiliary Equipment for Grate Firing

In grate- or stoker-fired boilers, the combustion of solid fuel occurs in a bed at the bottom of the furnace. Grate firing is the oldest method for the combustion of solid fuels. Numerous different applications of grate- or stoker-firing systems exist for the burning of different solid fuels. Biomass has been successfully burned with grates for hundreds of years. In the firing of solid biomass the fuel burns in some form of layer through which a large portion of the air for combustion passes. This primary air is forced through the grate and burning bed. The combustion process is controlled by the bed burning rate.

Grate firing was the main combustion technique for solid biomass up until the 1980s, when fluidized beds started to gain hold. For large boilers, grate firing has within the past 20 years largely been replaced by bubbling fluidized beds, but for under the 10 MWth range and for waste fuel firing, grate firing is still widely used.

Solid biomass size can be variable and there are no special requirements. The biomass has to be flowable. This means that the biomass is moved from the top of the grate to the bottom of the grate by gravity or by mechanical means. So the biomass needs to be chipped and, due to the combustion time, cannot contain large pieces.

Grate firing of solid biomass fuel does not need special auxiliary equipment although typically one needs ash handling, biomass handling, fans, blowers, and pumps. For biomass utilization one typically also has emission control equipment such as an electrostatic precipitator or fabric filter.

6.1.3   Auxiliary Equipment for Fluidized Bed Firing

Fluidized bed reactors have been used in petrochemistry and coal gasification since the 1930s. The first commercial applications for the combustion of solid fuels arrived in the 1970s (Basu, 2015). Bark and peat were among the first commercially accepted fuels. Since then, fluidized bed combustion has become widely accepted for the combustion of all kinds of solid fuels. Fluidized beds are very suitable for the combustion of low-grade fuels with a high moisture or ash content. These are normally difficult to burn using other combustion methods. The benefits of fluidized bed combustion are the possibility to use several different fuels simultaneously, simple and cheap sulfur removal by injecting limestone into the furnace, high combustion efficiency, and low NOx emissions.

The two main types of fluidized bed combustion are the bubbling fluidized bed boiler (BFB) and the circulating fluidized bed boiler (CFB). The BFB has lower own energy consumption and a lower unit price. The CFB has better environmental performance. Typically the BFB is used in applications of less than 100 MWth and the CFB from 50 MWth. The CFB is also better suited for coal burning.

Solid biomass is fed to fluidized bed boilers by screws or chutes (Basu, 2006). Therefore one needs to chip or reduce the size somewhat. Large biomass pieces do burn successfully in fluidized beds but the main problem is how to efficiently get them inside the furnace.

Fluidized bed firing of solid biomass fuel requires special auxiliary equipment for bottom ash handling and sand addition (Teir, 2004). Other typical auxiliary equipment includes fly ash handling, biomass handling, fans, blowers, and pumps. For biomass utilization one typically also has emission control equipment such as an electrostatic precipitator or fabric filter. If needed, a limestone addition to the furnace can be used for SO_2 removal.

6.2   Solid Fuel Handling

Solid fuels need to be stored in a safe manner and transported to the furnace. Often the size has to be changed for better burning properties. All equipment that participates in this work is called fuel-handling equipment.

Biofuels are often light, bridging, dusty, and moist and carry the risk of spontaneous combustion. Bridging means a biofuel will not flow out from a silo but supports its own weight from the walls. Because biofuels are bridging, the silo design is of high importance. Biofuels will become dusty and if the dust is dry it can ignite and cause a fire. Proper design of firefighting equipment, especially at conveyors, is important.

Biofuels tend to have impurities. Fuels that are bought sieved to less than 50 mm have been found to contain iron bars, bicycles, tree trunks, and stones of more than 0.5 m diameter. Reclassification of biofuel before it is transported from the silo is important.

6.2.1   Sizing and Removing Impurities

The most typical way of reducing the particle size of biomasses is chipping. In chipping, a sharp blade cuts biomass to 5–15-mm-thick pieces. Typically proper chipping results in a rather even size although there might be pins (i.e., long and narrow pencil-like pieces). Less typically one can reduce the particle size of biomasses by, e.g., hammermills or crushers. In this case the particle size is reduced to $\sim 100\,\mu m - 2\,mm$ for burning in suspension. The size is decreased by hitting the biomass with a rotating weight (hammer) or by crushing it between two uneven surfaces. Hammermills and crushers produce rather unevenly-sized particles. Stumps are often reduced in size by first cutting them into smaller pieces with hydraulic blades before crushing. Size reduction has been also tried for peat but no modern reference exists.

One can agree that the fuel brought to the plant is of a certain size. Magnetic separation can be used to remove iron-based impurities. Often, for cost reasons, magnetic separation is not used. Table or roll sizing is recommended to get rid of material that is too large. In spite of good intentions, biomasses often contain large rocks, pieces of tools, and different types of waste (Fig. 6.1).

6.2.2   Handling Biofuels

Biomass brought to the plant is first unloaded and then stored. A screw unloads biomass from the storage and drops it into a conveyor. The conveyor transports the biomass into the day silos beside the boiler. Day silos (6–18 hour storage) are

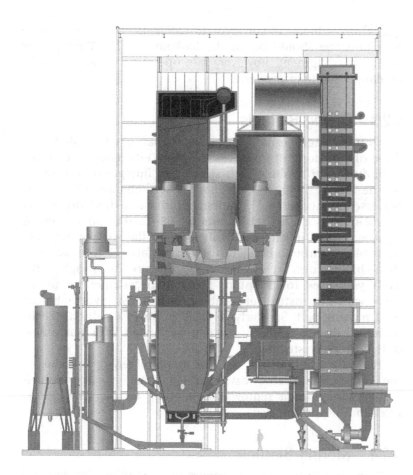

Figure 6.1 The handling of biomass feeding and ash in a fluidized boiler, Stora Enso Poland S.A., Ostroleka, Poland. Steam: 164 MWth, 62 kg/s, 114 bar, 520°C. Fuels: biomass, paper, and fiber rejects, sludge, coal. Courtesy of Valmet Power.

used to ensure continuous fuel flow to the boiler. Typically there are frequent problems and pauses in the biomass handling due to the irregular size of the biomass to be fired.

Handling biomass always presents a fire hazard. Almost every piece of equipment in contact with biomass can cause an explosion in specific conditions (Koppejan et al., 2013). Biomass handling creates fine dust. Fine dust dries easily. The fine dust layer can explode as it contains both dry combustible material with a high surface area and air.

Biomass stored in unventilated space will utilize all the air around it and replace it with carbon monoxide. If biomass is transported for a long time (e.g., in ships) the cargo holds should only be entered after proper ventilation and when one has ascertained that there is enough oxygen present.

Silos and conveyors should be equipped with fire alarms, cameras, water/steam spraying, safety exits, etc. It is very typical for a biomass conveyor to catch fire at some point in time and one should be on the lookout for this. With proper care and equipment, however, safe operation is possible with no danger to life or property.

6.2.3 Drying Biofuels

Biofuels often contain more water than dry fuel. Reducing the moisture content improves the plant efficiency. The drying of biofuels can be done with steam and flue gas dryers. Many of the older dryer models have been abandoned, mostly because of mechanical problems. Table and belt dryers are now widely used. In a table dryer a 0.05–0.10 m-thick biomass layer is pushed mechanically on top of a flat surface. Hot, dry air is blown through holes in the flat surface. Hot, moist air is then blown out, removing the moisture in roughly the same way used to dry pulp. The belt dryer operates in the same way, but instead of a table the biomass layer is placed on a moving belt. Similarly to table drying, hot, dry air is the preferred drying medium. Air can be heated with waste heat or low-pressure steam.

Typically one does not aim for a completely dry product – the target is 10–15% dryness. The heat consumption is 1.2–2 kWh/kg water removed, depending on the temperature level and the air flow used.

6.2.4 Storing Biofuels

Biomass can be stored outdoors in large piles. Weather and climate affect the fuel stored outside. In particular, snow and ice are problematic. Biomass has a low density, especially after it has been chipped. This means that storing substantial amounts often requires large storage facilities. Biomass can also be stored in silos. Silos are round structures made out of metal or concrete. Biomass will bridge. Bridging means that in spite of empty space below the biomass, it will not fall down because it forms a self-supporting arch. An igloo is an example of a self-supporting arch.

Moisture is the biggest problem for biomass storage. When biomass humidity is between 20% and 60%, it will begin to degrade. Biological degrading causes the local temperature to rise. If the local temperature increases too much, a fire may start with devastating consequences. Large biomass particles, such as wood chips, may form air pockets that have nearly ideal conditions for decomposing.

6.3 Liquid and Gaseous Fuel Firing

Burners are devices that produce heat though combustion, resulting in a visible flame. Burners can be divided into subcategories based on, e.g., fuel−air mixing:

a. Diffusion burners: Fuel and air are mixed by molecular diffusion. Thus the burning rate is controlled by diffusion.

b. Premixed burners: Fuel and air are partially mixed before the burner. Typically a portion that is much lower that the stoichiometric ratio is used. The reason for premixing is to increase combustion efficiency and decrease combustion time. Industrial examples are low calorific fuels (e.g., lignite and peat).

c. Kinetically controlled burners: Fuel and air mixing is controlled by aerodynamic and turbulence forces. Combustion is controlled purely by the form of the flame.

Burners are often sold by specialist companies and are often highly standardized due to long exposure to various rules and regulations pertaining to them.

6.3.1 Placement of Burners

The most typical burner arrangement for smaller boilers is the single wall arrangement. All burners are placed on one wall. This facilitates maintenance and operation costs. Because of the reduced layout and piping costs, this arrangement is economical. The most typical arrangement is the frontwall arrangement, Fig. 6.2.

When frontwall burners are fired, the flames in a large boiler can form an almost continuous flamesheet. When the required capacity increases, it becomes a more and more difficult engineering challenge to make large enough burners. Therefore when very large boilers are designed, it is advantageous to place burners on both the front- and the backwall. When placing large burners, it is customary to use horizontal spacing of 1.5−2.5 m and vertical spacing of 2−4 m.

Another option is to place burners on all four walls. The disadvantage of this is that the flames hit each other easily, leading to unstable combustion. An improved arrangement is to use corner firing, which increases mixing and facilitates turn-down. Corner firing is used especially in large coal-fired utility boilers but has also been used for biomass firing.

High moisture and low-grade fuels such as sludges and sulfite liquor require a long combustion time. Their adiabatic combustion temperature is low, requiring a refractory-lined

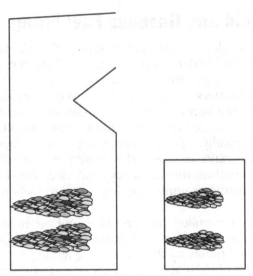

Figure 6.2 Frontwall burner arrangement, viewed from the top to the inside of the furnace.

combustion chamber. A typical burner application is roof or downshot firing, where a separate, often refractory-lined combustion chamber is built alongside the furnace proper. Fuel is then fired downwards in the separate combustion chamber before it enters the furnace proper.

6.3.2 Burner Design

The burner comprises of various parts. Typically fuel enters in the middle though a steel tube called a lance. Primary air is inserted from the center close to the fuel. Secondary air is often blown as another ring around the primary air. The burner is placed flush with the boiler wall. Wall tubes are bent to form an opening of suitable size, Fig. 6.3.

The burner needs additional parts. Ignition is typically done by an electrically ignited gas pilot flame. This small flame then ignites the main fuel. For safety reasons one needs a flame guard.

6.4 Air System

In boiler plants, forced draft (FD) fans supply primary, secondary, and tertiary air to the furnace. The air is primarily used for the combustion of fuels. Air can also be utilized for the pneumatic transport of fuels and other solid materials to

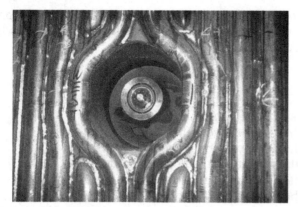

Figure 6.3 Gas burner at the furnace wall.

the furnace. In CFBs we need air to fluidize the return solids. Often we use air to seal sootblowing and fuel-handling equipment. The induced draft (ID) fans exhaust combustion gases from the boiler. In operation they normally produce a small underpressure in the furnace to make normal leakage occur from the working area to the furnace and not from the furnace to the working area. Fluidized bed boilers can be equipped with a flue gas recirculation fan. The recirculated flue gas is used for bed temperature and NOx control. Air and flue gas ducts must be gas tight and be able to endure over- and underpressure. This depends on the design pressure of the channel (at least ±5 kPa, sometimes +20 kPa). Flue gas ducts must be well isolated so the sulfur in the flue gases will not cause damage to the flue gas structures.

Flue gas velocity at maximum continuous rating (MCR) is chosen typically to 8–10 m/s at practical minimum load to prevent the accumulation of fly ash in the ducts. To reduce pressure losses and fan power consumption the flue gas velocities at full load must not exceed 30–35 m/s.

Performance curves provided by the fan manufacturer are typically used for the selection of the fan. The curves are based on design calculations and experiments that the manufacturer has made in their laboratories. These curves illustrate the change in the total pressure created by the fan as a function of volume flow and speed of rotation. When choosing a fan, the required volume flow and pressure difference must be known at several operating points. Other factors influencing choice are the fan's efficiency, the space required for installation and the shape of the characteristic curve for the fan.

The following formula provides a calculation for the pressure losses caused by flow resistance when air or flue gases flow in the boiler channels:

$$\Delta p = \frac{\rho w^2}{2}\left(\xi\frac{l}{d} + \sum \zeta_i\right) \qquad (6.1)$$

where

l is the channel length, m
d is the channel hydraulic diameter, m
ρ is the density of the flowing matter, kg/m^3
w is the velocity of the flowing matter, m/s
ζ is the single resistance coefficient, -
ξ is the friction coefficient, -

Sudden changes in direction, narrowings, and enlargements must be avoided in order to minimize the pressure losses in ducts. This allows minimization of the single resistance coefficient in the above formula.

A typical pressure profile in the boiler has two sides. Fans draw air and pressurize it. The FD air fan controls the air flow and thus the oxygen content. The primary air flow through the fluidized bed or the grate causes the greatest pressure loss up to 10–15,000 Pa. Often the secondary and tertiary air fans provide much less pressure. The furnace operates at 100–200 Pa underpressure. Flue gas flow through the dense tube bundles in heat exchangers further reduces the pressure. The ID fan controls the furnace underpressure. The stack causes a natural draft, which reduces the required static head of the ID fan. A natural draft is caused by the density difference of the hot, light flue gas inside the stack and the colder, denser air outside the stack.

6.4.1 Air Fans

Most fans in boilers are the radial type. Sometimes a two-sided air inlet is preferred (Fig. 6.4). Axial fans are usually avoided as they are more expensive. In a radial fan the rotating blades in the rotor accelerate the flow out from them. This velocity is then converted to pressure when the flow decelerates. As velocities much over 100 m/s are hard to achieve, the maximum pressure in a radial fan is limited.

6.4.2 Steam Air Preheaters

Steam air preheaters are used to preheat the air flow to the furnace. Steam flows from one end of the tubes to the inside tubes. Air flows outside these finned or bare tubes and is heated by the condensing steam (Fig. 6.5). Condensate flows out from

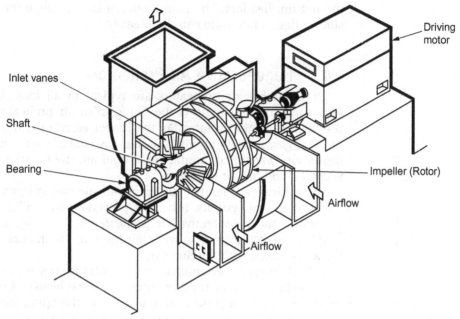

Figure 6.4 Radial air fan. From Singer, J.G. (ed.), 1991. Combustion Fossil Power, fourth ed. Asea Brown Boveri, 977 p. ISBN 0960597409 (Singer, 1991).

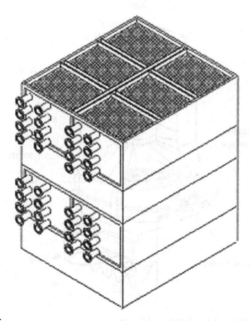

Figure 6.5 Steam air preheater.

the bottom headers. The main concern is to arrange the steam side so that condensate can drain easily.

6.4.3 Regenerative Air Preheaters

Regenerative air preheaters are typical in all large boilers. The most common type is the Ljungström air preheater. In it the heater element rotates. As the heater element contacts the hot flue gas, it heats up and the flue gas cools. Then, when the heater element makes contact with cold air, the heater element is cooled and the air is heated.

In an ideal regenerative air preheater the flue gas and air do not mix. In real systems, mixing is prevented by sealing plates. Sealing of the regenerative air preheater is, however, a major problem. Rotating tends to wear out the seal. In practice 10% of the air is mixed with the flue gas.

A Rothemühle regenerative air preheater is shown in Fig. 6.6. In it the heater element cover rotates and the heater element is fixed. The design was developed to counter the Ljungström patent, where the heater element rotates and flows are fixed.

A typical choice for the regenerative air heater material is either an enamel-coated ceramic element or a metallic dimple

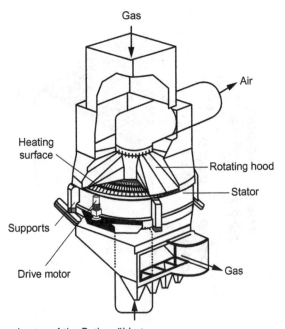

Figure 6.6 Regenerative air preheater of the Rothemühle type.

element. A corrosion-protective coating is popular because low-temperature corrosion is fairly typical. Metallic elements have higher efficiency than ceramic, require lower height, and have a lower pressure drop. The problem with metallic elements is the high corrosion rate in some applications.

6.4.4 Fluidized Bed Air Nozzles

To distribute air in the fluidized bed, special nozzles are needed. The nozzles try to keep the bed material from entering the air system. Primary air is introduced from the fluidized bed bottom using special nozzles. This air is used, in addition to combustion, for fluidizing bed particles.

Primary air is delivered either through a cooled membrane wall bottom, Fig. 6.7, or through nozzle arrays connected to cooled beams. Replaceable nozzles are used because the erosion on the lower part of the furnace is high. Most typical nozzle types are S-type and Cap-type.

Fluidization air is delivered through cooled membrane wall bottom, Fig. 6.7. Replaceable nozzles are used because the sand wear on lower furnace is high. Most typical nozzle types are S-type and Cap-type.

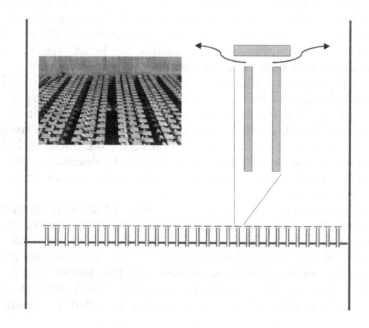

Figure 6.7 Air nozzles in the bottom of a fluidized bed.

6.5 Sootblowing

Heat transfer surface fouling is significant when we burn biomass with a high ash content. Due to deposits, the boiler thermal efficiency is reduced and corrosion of the heat surfaces occurs faster. This can result in increased boiler downtime. For this reason the heating surfaces should be cleaned constantly. Thorough cleaning of the boiler requires maintenance outage and washing with water, which includes cleaning the boiler heating surfaces with hot water (Pophali et al., 2013; Jameel et al., 1994).

Heat transfer surfaces can be cleaned manually in case of emergency, e.g., by the so-called chill-and-blow technique. In this the liquor combustion is stopped and the boiler is run merely by auxiliary fuel. The change caused by the decrease in flue gas temperatures and the resulting thermal contraction removes deposits from the surface of the tubes. This is not a suitable method for the continuous cleaning of heat surfaces because of the high strain and process disturbances it causes (Vakkilainen, 2005).

6.5.1 Steam Sootblower Design

In practice, recovery boiler surface cleaning is done with the help of steam sootblowers. The most typical sootblower has a rotating lance, which is inserted from the wall to clean inside the heat exchanger surface, Fig. 6.8. This type of sootblower slowly rotates as it is inserted and retracted. When not in use the sootblower is pulled outside the boiler. The tip of the lance has a hole. Steam at sonic speed is injected from the tip. The steam jet then removes the deposit from the heat transfer surfaces either by shearing or by breaking the deposit into pieces. The steam jet also causes a heat shock when the deposit is knocked off. Typically sootblowers in the past used steam at reduced pressure after the primary superheaters, but now many boilers use steam extraction of the desired pressure from a steam turbine. Typically, sootblowers use 18–25 bar pressure steam.

A steam sootblower can clean the deposit from heat transfer surfaces by means of two mechanisms. It either breaks the deposit into small pieces or pushes the deposit off the surface of the tube as large chunks. Both of these mechanisms are based on breaking the weakest bonds in the deposit and the ability of the steam jet power to exceed the tensile strength of the deposit. Deposit removal as pieces is called the brittle

Figure 6.8 Retractable sootblower.

break-up mechanism, and deposit removal from surfaces is called the debonding mechanism. Which mechanism takes place in the cleaning process depends on the properties of the deposit as well as on the place where the jet hits the tube plates (Pophali et al., 2013).

6.5.2 Deposit Removal

The two essential features of deposit removal are tensile strength and adhesive strength. Tensile strength indicates how much power is needed to break the deposit into pieces, and adhesive strength describes how firmly the ash deposit layer is attached to the tube surface.

For optimal functioning of the sootblower, as much as possible of the steam pressure should be converted into jet kinetic energy. High-jet kinetic energy helps to break or debond hard-to-remove deposit from heat transfer surfaces. Typically the steam mass flow rate through the nozzle is directly proportional to the driving steam pressure. The steam mass flow rate through the nozzle can vary from 1.5–2.5 kg/s. Sootblowing steam consumption is thus significant in this instance (Fig. 6.9).

A measure of the sootblowing jet cleaning power is the peak impact pressure (PIP). PIP is defined as

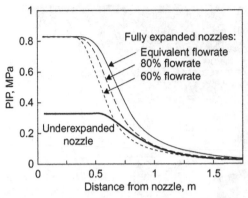

Figure 6.9 Decline of exit velocity of supersonic turbulent sootblower jet. From Kaliazine, A., Cormack, D.E., Eibrahimi-Sabet, A., Tran, H., 1998. The meachanics of deposit removal in Kraft recovery boilers. Proceedings of International Chemical Recovery Conference. 1−4 June 1998, vol. 3. TAPPI, Tampa, USA, pp. 641−654 (Kaliazine et al., 1998).

$$PIP = \frac{1}{2}\rho w^2 \qquad (6.2)$$

where ρ is the vapor density, and w is the jet speed. Pressure generated by the sootblower drops sharply with increasing distance from the sootblower nozzle because the jet spreads and slows down. One meter away from the nozzle, the PIP is about a decade less than at the inlet. Fig. 6.9 presents the PIP decline as a function of distance for different nozzles (Kaliazine et al., 1998). As can be seen, the effective cleaning radius of a sootblower jet is 1−1.5 m.

6.6 Dust Removal From Flue Gas

Biomasses contain ash. Typically the largest portion of ash can be removed from the bottom of the furnace, but small ash particles, called fly ash, get carried away with the flue gas, and need to be removed. Some of the fuel does not burn and shows up as unburned carbon in the fly ash. Fly ash also contains deposition from the condensable material in the flue gas. Dust has long been considered a nuisance, so as much of it needs to be removed as possible. Depending on the fuel and the boiler type, there is variation in the particle size, composition, and the amount of dust in the flue gas. Typical devices to remove dust are electrostatic precipitators, filters, and scrubbers (Fig. 6.10).

Figure 6.10 Electrostatic precipitator being erected.

6.6.1 Electrostatic Precipitators

An electrostatic precipitator's efficiency depends on its area, the flue gas flow, and the electrical properties of the dust.

$$\eta = 1 - e^{-\left(\frac{Aw}{V}\right)^{k}} \qquad (6.3)$$

e is the collection efficiency, -
A is the collector surface, m^2
w is the collection migration velocity, m/s
V is the actual gas flow, m^3/s
k is an empirical constant (0.4−0.5−0.6), -

The principle behind the electrostatic precipitator is simple. Dust-laden flue gas flows along collecting surfaces formed of plates. Strings of wire are suspended between the platens. When high-voltage electric current is applied between the electrodes (−) and the collector plates (+), the dust particles become charged. The charge draws them towards the collecting plates, where they deposit.

6.6.2 Fabric Filters

A fabric filter is basically a porous substance. When the flue gas flows through it, those particles, which are larger than the pores, are collected. Collected ash particles form a deposit.

This acts as an additional separation surface. The benefit of a fabric filter is that, metals, e.g., in the flue gas can deposit on collected dust.

Deposit removal typically occurs by means of a pressure wave. The layer thickness is monitored through pressure difference or timing. When it gets high enough, one initiates a cleaning cycle, which causes the deposit on the dirty side to peel off.

Large fabric filters are often constructed of tubular surfaces arranged so that flow is from the outside of the tube to the inside of the tube. Filter cleaning can be done by a pulse jet, reversed air, or mechanical vibration. In pulse jet cleaning the cleaning action is caused by a short air blast that travels along the fabric cloth, shaking the dust from the cloth surface. In reverse air cleaning, the filtering action is stopped and a flow of air is sent in the opposite direction from that of normal operation, and this separates the dust layer from the cloth. In small filters the dust separation can be done by mechanically displacing the top of the filter cloth tube. Side movement causes the deposits to be removed. Mechanical devices are the least favored as their use often shortens the lifetime of the fabric material.

The drawback of the fabric filter is that the operating temperature is restricted. Few of the surface materials used can be safely operated for a long time at temperatures close to 200°C. Changing the filter cloth during operation is generally not possible, so one waits until the operation deteriorates before scheduling a shutdown. A particular problem with fabric filters is the ash adhesion to the cloth caused by the flue gas dew point.

6.7   Ash Handling

Ash is the common name for all combustion residues. Ash must be sootblown from surfaces. Then the dust in the flue gas must be separated with, e.g., an electrostatic precipitator and transported by conveyors or pneumatic systems (Fig. 6.11).

6.7.1   Submerged Ash Conveying

In fluidized bed boilers the ash is taken out continually from the bottom of the furnace. Conveyors cooled with water are frequently used. Air is used to cool the ash system closer to the furnace. Cooled ash is dragged for removal by a bottom conveyor (Fig. 6.12).

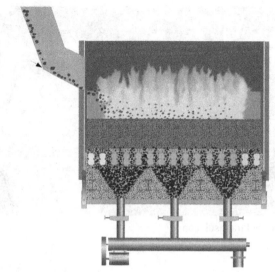

Figure 6.11 Bottom ash system.

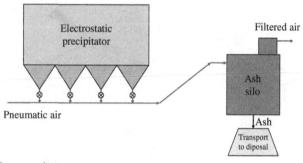

Figure 6.12 Pneumatic ash conveying.

6.7.2 Pneumatic Ash Conveying

In pneumatic ash conveying, pressurized air is used to blow ash from one place to another. One typical example is conveying electrostatic precipitator ash to a silo. Pneumatic ash conveying requires a pressurized air flow.

Pneumatic ash conveying is used when ash flow is moderate (e.g., in peat boilers). Ash must also preferably be uniform in size and of small diameter.

6.7.3 Ash Conveyors

Ash is collected in hoppers. Motors turn the ash conveyors. Collected ash is dropped into ash chutes (Fig. 6.13).

Fig. 6.14 shows a section of an ash conveyor. Ash is dropped into the conveyor from the top. It is dragged along the bottom

Figure 6.13 Typical ash conveyor before insulation.

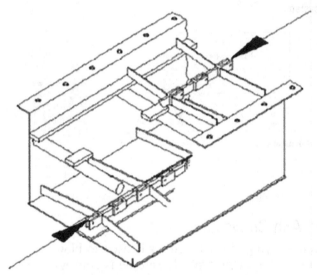

Figure 6.14 Section of an ash conveyor.

to the right with moving scraper blades. These are attached to a belt. The belt returns to the top of the conveyor. The return leg can also be below the conveyor bottom.

6.8 Silencer

During safety valve operation, high-pressure steam expands into the atmosphere. This steam has a speed close to sonic speed. Without a silencer the noise can be heard far away (Fig. 6.15).

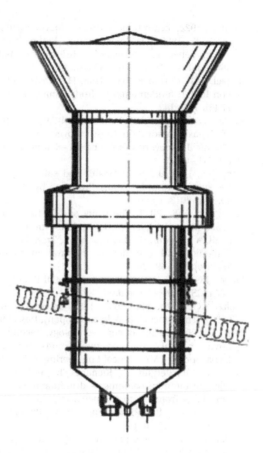

Figure 6.15 Silencer.

References

Akturk, N.U., Allan, R.E., Barrett, A.A., Brooks, W.J.D., Cooper, J.R.P., Harris, C.P., et al., 1991. In: Clapp, R.M. (Ed.), Modern Power Station Practice, vol. B. Boilers and Ancillary Plant, third ed. Pergamon press, Singapore, 184 p. ISBN 0080405126.

Basu, P., 2006. Combustion and Gasification in Fluidized Beds. CRC Press, 345 p. ISBN 0849333962.

Basu, P., 2015. Circulating Fluidized Bed Boilers: Design, Operation and Maintenance. Springer, 366 p. ISBN 9783319061726.

Basu, P., Fraser, S.A., 1991. Circulating Fluidized Bed Boilers. CRC Press, 345 p. ISBN 075069226X.

Bergman, P.C.A., Boersma, A.R., Zwart, R.W.R., Kiel, J.H.A., 2005. Torrefaction for biomass co-firing in existing coal-fired power stations. ECN Report, ECN-C—05-180, July 2005, 71 p.

Brindle, M.J., Cresswell, I., Greenslade, J.C., Jones, A.R., Jones, K., Martin, P.C., et al., 1991. In: Martin, P.C., Hannah, I.W. (Eds.), Modern Power Station Practice, vol. A. Station planning and design, third ed. Pergamon press, Singapore. ISBN 0080405118.

Croft, T., 1922. Steam Boilers, first ed. McGraw-Hill, New York, 412 p.

Doležal, R. 1985. Dampferzeugung: Verbrennung, feuerung, dampferzeuger (Steam generation, combustion, firing, steam boilers). Springer Verlag, Berlin, 362 p. ISBN: 3540137718 (in German).

Jameel, M.I., Cormack, D.E., Tran, H., Moskal, T.E., 1994. Sootblower optimization Part 1: fundamental hydrodynamics of a sootblower nozzle jet. TAPPI J. 77 (5), 135–142.

Kaliazine, A., Cormack, D.E., Eibrahimi-Sabet, A., Tran, H., 1998. The meachanics of deposit removal in Kraft recovery boilers. Proceedings of International Chemical Recovery Conference. 1–4 June 1998, vol. 3. TAPPI, Tampa, USA, pp. 641–654.

Koppejan, J. et al., 2013. Health and safety aspects of solid biomass storage, transportation and feeding. Produced by IEA Bioenergy Task 32, 36, 37 and 40, May 2013, 100 p.

Livingston, WR., 2013. The firing and co-firing of biomass in large pulverised coal boilers. IEA Exco Workshop Jeju, November 20131, 20 p.

Pophali, A., Emami, B., Bussmann, M., Tran, H., 2013. Studies on sootblower jet dynamics and ash deposit removal in industrial boilers. Fuel Process. Technol. 105, 69–76.

Singer, J.G. (ed.), 1991. Combustion Fossil Power, fourth ed. Asea Brown Boveri, 977 p. ISBN 0960597409.

Stultz, S.C., Kitto, J.B. (Eds.), 1992. Steam Its Generation and Use, 40th ed. 929 p., The Babcock & Wilcox Company. ISBN 0963457004.

Teir, S., 2004. Steam Boiler Technology, second ed. Energy Engineering and Environmental Protection publications, Helsinki University of Technology, Department of Mechanical Engineering. 215 p. ISBN 9512267594.

Wadsborn, R., Berglin, N., Richards, T., 2007. Konvertering av mesaugnar från olje- till biobränsleeldning – drifterfarenheter och modellering. (Converting lime kilns from oil to biofuel firing) In Swedish Rapport, Skogsindustriella programmet, Rapportnummer 1040, Projektnummer S5-606, Värmeforsk, December 2007, 48 p.

Vakkilainen, E.K., 2005. Kraft recovery boilers—principles and practice. Suomen Soodakattilayhdistys r.y., Valopaino Oy. Helsinki, Finland, 246 p. ISBN 9529186037.

7

BOILER MECHANICAL DESIGN

Boiler mechanical design deals with the design of boiler parts. In addition to the design of pressure parts, one must design the mechanical structure. Boiler structures have changed a lot over the years. However, many old designs are still used today. Pressure part design has a lot to do with pressure part manufacture. That is to say, the target is always the lowest manufacturing cost (Advances, 1986; Benesch, 2001).

All large boilers are hung from top beams because of thermal stresses (Combustion fossil power, 1991). Boiler buildings are almost always made using a steel structure. Platforms are used for working access to different parts of the boiler. The pressure part is insulated to decrease the heat losses.

7.1   Furnace Walls

Modern furnaces are made gas tight by welding tubes to form membrane walls, Fig. 7.1. All surfaces (walls, bottom) are

Steam Generation from Biomass. DOI: http://dx.doi.org/10.1016/B978-0-12-804389-9.00007-1

Figure 7.1 Modern furnace walls with air ports.

cooled by circulating a steam–water mixture. There are openings to introduce air and fuel to the furnace and to extract ash from the furnace. The lower part of fluidized bed furnaces are lined with refractory. Refractory is held in place by anchors, which are welded to the membrane wall fins. Air ports, openings, etc. will be filled with refractory before the hydro test, providing that there are no tube butt-weld joints inside the box. Most refractory filling will already have been done in a workshop, but if the case the box has been split for transportation because of the size of the components, the filling must be done on site.

Openings such as access doors, burner openings, measurement openings, and air ports are made by bending tubes (Doležal, 1967). Gas tightness of large openings is ensured by means of refractory and plate. Room must be left for insulation. Cast iron doors and refractory blocks are used to close larger openings when they are not in use, Fig. 7.2.

The membrane wall is of universal design in modern water-tube boilers (Doležal, 1985). A gas-tight wall is created by welding fins between tubes, Fig. 7.3. The fin's width is usually much smaller than the tube's outer diameter. In furnaces, fins of 10–20 mm are typically used. In the backpass convection sections, fins of 50–80 mm can be used. Each fin is welded to tubing. The most typical construction uses four welds to connect a fin between two tubes. Because there are two passes, the welds can be smaller. As tubes are welded from both sides, it is easy to achieve symmetry. When the boiler is constructed, wall

Figure 7.2 Temperature measurement and inspection openings in a furnace wall.

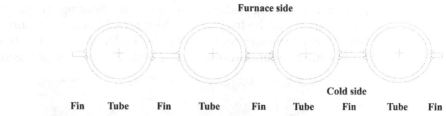

Figure 7.3 Membrane wall.

elements comprising some tens of tubes are joined together. It is easy to make these joining field welds if half fins have been welded to tubes. This also ensures that during field joining, pressure part tubes are not accidently damaged.

The first gas-tight construction involved placing a steel plate behind the tubes (Ledinegg, 1966). Insulation was rigged on top of the plate. At first, tubes were not connected to each other. This proved to be a problem as small spaces transfer corrosive gases and larger spaces transfer heat. Later it was common to make gas-tight walls by placing tubes alongside each other and welding them together. This construction is cheaper than the finned construction. However, it is not used much now because tube leaks in this kind of wall are very difficult to repair. Fins were also made by welding smaller pieces to the tube and then joining these together. This practice has also stopped.

Especially in the 1950s and 1960s there was a tendency to use specially drawn tubes. Tubes can be drawn to have fins in them. Then the only remaining thing is to join these tubes together. The price of these special tubes has made this kind of structure uneconomical except for special heat transfer surfaces.

7.2 Superheaters

Superheaters are generally placed where the flue gases are hot. There is still a lot of molten ash and reactive gases. This places an extra burden on the mechanical construction of superheaters (Gaffert, 1981). Because superheaters have to be built for optimal behavior in an often hostile environment that differs from boiler to boiler, there are very many different types of superheater (Fig. 7.4).

7.2.1 Radiative Superheater

Radiative superheaters are used as the first heat transfer surfaces after the furnace. Often a platen arrangement is preferred to create a fouling-resistant construction. Spacing in the first superheaters can be from 150 to 1000 mm. Superheaters are the hottest heat transfer surfaces in the boiler, so they are often built with temperature-resistant alloys. As these superheaters

Figure 7.4 Superheaters under construction. Courtesy of Andritz.

operate at flue gas temperature ranges of 800–1200°C, they often need to be built with corrosion resistance in mind. High-grade materials are expensive and require special welding techniques and often heat treatment. Superheater type, design, material, and construction methods always involve compromise to achieve the desired technical level at low cost.

The wall superheater is a special class of superheater. Sometimes part of the superheating surface needs to be located at the furnace proper. The wall superheater surface is used in these applications. A wall superheater is basically a flat panel of joined tubes that hangs either as close to the furnace membrane wall as possible or forms part of the furnace. Sometimes the backpass side walls form a superheating surface. Typically wall superheaters are the first superheater after the steam drum or the coolest superheater.

7.2.2 Convective Superheater

When flue gas temperatures decrease, the radiative flux gets smaller and the share of convective heat transfer is increased. If less than half of the heat transfer is from radiation, the superheater is designated as a convective superheater. Because of the flue gas temperature decrease, the ash is no longer so sticky, and the tubes in superheating surfaces can be placed closer to each other. The flue gas velocities can be increased. Transversal spacings of even less than 100 mm can be used. Convective superheaters are often placed in the backpass with the flue gas flow downwards. As tube temperatures are high, the tube material needs to be high-grade steel with an appropriate amount of Cr and Ni to ensure adequate strength at operating temperatures.

7.3 Economizer

The economizer has the lowest steel temperature of any heat exchanger surface in the pressure vessel. Therefore it can usually be constructed from carbon steel (Teir, 2004). The economizer is placed as the last of the flue gas to steam–water heat exchanges. The flue gas flow has a low temperature, so fouling is normally not a big problem. Therefore both the transversal spacing and longitudinal spacing can be tight. Economizers are almost without exception countercurrent heat transfer surfaces. If the flue gases are cooled below 150°C, the low temperature corrosion can result in restrictions to the economizer design and construction.

The most common economizer has carbon steel tubes from 30 to 50 mm outer diameter. Transversal and longitudinal spacing is almost the same, from 50 to 150 mm. Because of low temperatures, the heat transfer coefficient in economizers tends to be small. To increase the heat transfer surface the economizer tubes can be made with fins. In biomass applications the finned heat transfer surfaces tend to become fouled and are seldom used.

7.4 Air Preheaters

In air preheaters the air is heated with flue gas or steam to a higher temperature. Heating combustion air increases the furnace temperatures and improves boiler operation. Air preheaters often have the largest surface area of all heat exchange surfaces. This is because heat transfer coefficients from gas to gas are very low. Air preheaters can be constructed from carbon steel but corrosion at low temperature is always a problem.

7.4.1 Recuperative Air Preheaters

In recuperative air preheating the heat flows from higher temperature media (flue gas through the heat exchanger walls) to lower temperature media (air). This is the most typical heat transfer surface in biomass boilers.

Because the stresses are low, the wall thicknesses are only a few millimeters. With thin and fairly big tubes one needs to be careful to eliminate vibration. As the air flow is large and velocities cannot be too high due to too high own power, there are often around a hundred parallel tubes. This means that there are often only two to three passes in each air preheater.

As a large part of the pressure loss is from distribution, turning, and exit rather than from friction losses, the dimensioning of ducts to and from the air preheater—and especially the turns from one pass to another—require careful analysis.

Recuperative air preheater walls are made from steel plate with insulation. The walls need to be strengthened to withstand pressure swings. A large part of the manufacturing cost is fitting tube ends to these steel plates.

7.4.2 Regenerative Air Preheaters

Regenerative air heaters are rotary devices. In regenerative air heaters the outgoing flue gas heats mass that is slowly

Figure 7.5 Ljungström-type air preheater.

rotated through the flue gas flow. Flue gas heat is thus stored in the rotating material (Stultz and Kitto, 1992). When then rotating material is moved to the incoming air side, the heat is transferred from hot material to colder air (Fig. 7.5).

7.5 Boiler Pressure Vessel Manufacture

There are two main work methods used in modern boiler pressure vessel manufacture. Welding is used in the manufacture of membrane walls and for joining tubes together. Plate and tube rolling is used to make the steam drum, form expanding or contracting tube pieces and to form headers. Iron can be joined by melting similar material either with hot gas or an electric arc. The welding procedure depends on the material and manufacture.

When doing tube panel welding, the most typical construction is to use four welds to connect a fin between two tubes. Because there are two passes (i.e., welding from both the top and the bottom direction), the welds can be smaller. If tubes are welded simultaneously from both sides, it is easy to achieve symmetry.

7.6 Pressure Part Design

The design of pressure parts is controlled by applicable laws, codes, and instructions. To build, erect, and design pressure vessels one needs a permit. The most important factors are design procedure and materials (Fryling, 1967).

In the early 1800s there were several notorious instances in the United States involving explosions in steamboat boilers, which had the largest capacity of steam boilers. This led to the passing of Acts in 1838 and 1852 requiring licensing and periodic inspection of boilers and establishing a maximum operating steam pressure. This act markedly decreased the number of steamboat accidents and fatalities (Hunter, 1985).

There are several pressure vessel codes used in various countries. Previously, each country used to have its own pressure vessel code and its own pressure vessel design rules. Of the country codes the ones of the American Society of Mechanical Engineers (ASME) and German Deutsches Institut für Normung (DIN) were most often used. Now DIN has been surpassed by the European Union-wide European Standards (EN), which are used in addition to Europe widely in Asia. The ASME pressure vessel code continues to be accepted in numerous countries around the globe. The use of other national codes, such as Finnish Standards (SFS), British Standards (BS), and Russian gosudarstvennyy standart (Gost), is gradually fading.

To be able to show proficiency, each party in steam boiler manufacture must have a permit from an authority to use such a code.

7.7 Pressure Part Materials

One of the most important choices in pressure part design is to choose the right material. When material is chosen, one must look at operating conditions (pressure, temperature, corrosion, erosion), mechanical properties (strength, ductility), manufacturing properties (weldability, heat treatment requirements), cost of material, and manufacture and availability (supply time).

Typically, pressure parts are made from different grades of steel. The properties of the materials depend on their composition and also on the type of structure the steel has. Structure can be changed by, e.g., heating and cooling the material.

7.7.1 Boiler Tube Materials

Boiler tubes are divided into several classes. Examples of common boiler tube materials are shown in Table 7.1. Carbon steel tubes contain mostly iron. These are universally used in all pressure parts up to 400−450°C. Furnaces, economizers, drums, and connecting piping are usually made from common carbon

Table 7.1 Examples of Common Boiler Tube Materials

Class	Temperature (°C)	Typical Material (Common Name)	EN Number	ASME
Steel	Up to 400–450	P235GH, P265GH	EN1.0425	SA/A-210 Gr A-1
Alloyed	Up to 540–580	16Mo3, 13CrMo4-5, 10CrMo9-10	EN1.5415, EN1.7335, EN1.7380	SA/A-204, SA/A-387 12, SA/A-387 22
Martensite	Up to 550–600	X10CrMoVNb9-1, X20CrMoV11-1	EN1.4903	
Austenitic	Up to 600–720	X10CrNiMo1613, X10CrNiNb1613		

steel. Alloyed tubes are steel tubes with chrome up to 15 w-% with molybdenum and other additives. Alloyed tubes can be used at higher temperatures than carbon steel ones. These types of steel find a use in superheaters and can be used in most common boilers.

Martensite tubes have higher amounts of compounds other than steel. Usually their welding is difficult, so they are used only in limited high-temperature applications. Austenitic tubes are the most alloyed of all tubes. Their application is in tubes that require high temperature and high corrosion resistance. Austenitic tubes are prone to water (steam) side corrosion. Often in biomass application one uses even more alloyed tubes for corrosion protection.

Typically the properties for common tube and pipe materials can be found in standards. Material manufacturers can then supply these materials with certificates indicating that they fulfill the standard requirements. Seemingly similar materials can have very different allowable stresses. This is caused mainly by different quality control requirements and the level of allowable impurities. Materials with less quality control and higher levels of impurity are more prone to material defects.

Some standards divide material strengths based on the type of product: thin-walled tubes, thick-walled tubes, plate, forgings, and castings. This difference is caused by differences in manufacturing methods. All properties should be measured at the right temperature. For example, material properties at room

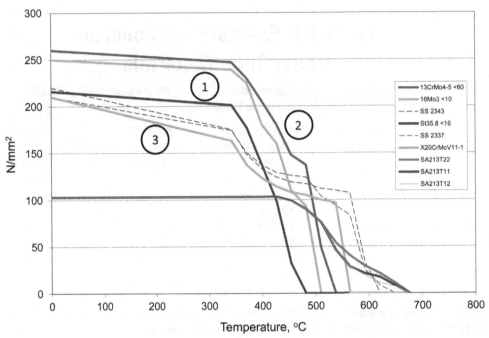

Figure 7.6 Allowable stresses for some common boiler tubes: 1, normal iron; 2, alloyed steel; 3, steel for high temperature.

temperature cannot be used to indicate material properties at high temperatures. The most important material property for pressure part design is the allowable stress, and in steam generators especially the allowable stress at high temperatures (Fig. 7.6).

Allowable stresses that can be used when designing the pressure part structure vary with the type of steel. Normal carbon steel has a high allowable stress at room temperature but a poor temperature range. After about 400°C the carbon steel has practically no allowable stress. Alloyed steel gives the designer the highest allowable stress at room temperature and increases the temperature range for these types of steel. The downside is the higher cost of these materials. High temperature steel is designed specifically to give at least some allowable stress at the highest temperatures (Fig. 7.7).

When designing a pressure vessel structure, appropriate allowable strength must be determined. This strength is usually found in the code. The weakest link determines the strength. Depending on temperature range, this is either tensile strength, yield strength, or creep rate.

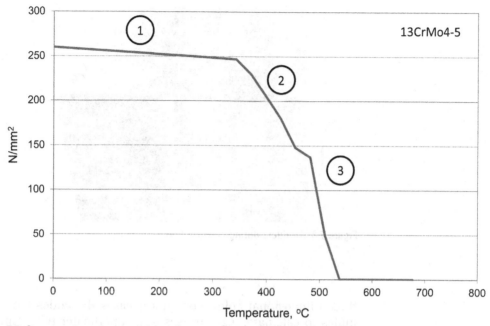

Figure 7.7 Stress regimes for steel: 1, tensile strength; 2, yield strength: 3, creep rate.

Steam boilers are normally constructed from ductile materials. Ductility can be defined by the amount of elongation before the material yields. Normal pressure part materials have elongation before rupture in the order of 20–25%. The use of standard methods of stress calculation requires ductile materials as ductility ensures even stress distribution.

Hardness is a surface property of the material. Hard surfaces are resistant to scratching, abrasion, and denting. Hard materials are also usually difficult to cut and drill. Hardness is usually measured with indentation apparatus. A specified pressure is applied to the surface for a specified time using specified apparatus. The indentation at the surface is then measured. Hardness is a very informative test to indicate differences in local properties caused by heat, corrosion, or manufacturing methods.

7.7.2 Stress Analysis

The stresses in a structure can be determined by using a suitable stress analysis program. The structure is divided into small segments. The loads and thermal stresses are added and the stress level in each node can be calculated and shown in

Figure 7.8 Air ducts during construction.

three-dimensional color. For typical cases the codes have formulas to calculate, e.g., stresses in a superheater tube. Current practice is to use stress analysis programs to study all "nonstandard" designs. Current codes are very dependable as they have years and years of use. Determining the right load case is the main problem.

The reduction of thermal stresses is one of the key issues in steam boiler design. Imbalances in water/steam/flue gas flow result in different temperatures. Differences in temperatures cause thermal stresses. Proper design must be made to take these into account. The key is flexibility and bending of the tubes (Effenberger, 2000).

Pressure parts should also be designed to withstand appropriate extra loads. These additional loads can be wind load, pressure shocks (water hammer), earthquake, snow load, and loading from interior material (ash buildup).

7.8 Ducts

Air ducts are usually made from carbon steel plate (Fig. 7.8). Bars are used to stiffen the structure. Ducts hang from supports. If the media inside has a high temperature, then insulation is used (Fig. 7.9).

Expansion joints need to be used to counter thermal expansion. Joints are mostly of fabric type. High temperature or corrosion requires metallic-type joints.

Figure 7.9 Expansion joints in an air duct.

References

Advances in power station construction, 1986. Central Electricity Generating Board. UK, Pergamon Press, Barnwood, Gloucester, 759 p. ISBN 0080316778.

Benesch, W.A., 2001. Planning new coal-fired power plants. VGB Power Tech. 81 (6), 37−44.

Singer, J.G., (Ed.), 1991. Combustion Fossil Power, fourth ed. Asea Brown Boveri, 977 p. ISBN 0960597409.

Doležal, R., 1967. Large Boiler Furnaces. Elsevier Publishing Company, 394 p.

Doležal, R., 1985. Dampferzeugung: Verbrennung, feuerung, dampferzeuger (Steam generation, combustion, firing, steam boilers). Springer Verlag, Berlin, 362 p. ISBN 3540137718 (in German).

Effenberger, H., 2000. Dampferzeugung (Steam boilers). Springer Verlag, Berlin, 852 p. ISBN 3540641750 (in German).

Fryling, G.R., 1967. Combustion Engineering, A Reference Book on Fuel Burning and Steam Generation. Second ed. The Riverside Press, Cambridge, Massachusetts, 831 p.

Gaffert, G.A., 1981. Centrales de vapor (Steam power stations). Editorial reverté, S. A., Barcelona, 602 p. ISBN 8429148302 (in Spanish).

Hunter, L.C., 1985. A History of Industrial Power in the United States 1780−1930 : Steam Power, vol. 2. University Press of Virginia, Charlotteville, 732 p. ISBN 0813907829.

Ledinegg, M., 1966. Dampferzeugung dampfkessel, feuerungen. (Steam power, steam boiler, firing). Springer−Verlag, Wien, 485 p.

Stultz, S.C., Kitto, J.B., (Eds.), 1992. Steam Its Generation and Use, 40th ed. 929 p. ISBN 0963457004.

Teir, S., 2004. Steam Boiler Technology. 2nd ed. Energy Engineering and Environmental Protection publications, Helsinki University of Technology, Department of Mechanical Engineering, 215 p. ISBN 9512267594.

8

AVAILABILITY AND RELIABILITY

From a user's point of view a steam generator is an expensive piece of equipment. If for some reason it cannot be used to generate steam, a great deal of money is lost. It pays, therefore, to invest in proper materials and maintenance to keep the

Steam Generation from Biomass. DOI: http://dx.doi.org/10.1016/B978-0-12-804389-9.00008-3

machinery operating. Thus steam boiler availability and safety have long been of utmost importance (Parrot, 1829). In Chapter 7, Boiler Mechanical Design, heat transfer surface sizing was discussed. It was pointed out that fouling requires an increase in the heat transfer surface area. So, what is fouling of heat transfer surfaces and how can it be reduced?

To maintain boiler operation one has to manufacture pure enough water to fulfill the requirement. Boiler water treatment makes water pure enough to operate the boiler safely. The most problematic aspect of boiler operation is finding the extent of the corrosion of heat transfer surfaces or the erosion of heat transfer surfaces.

8.1 Availability

When we look at the boiler operation in Fig. 8.1, we note that the boiler has produced steam less than 100% of the time. The time during which the boiler should have been able to produce steam, but did not, we call unavailability. Some events causing unavailability were planned in advance. Such planned unavailability could have been, e.g., a stoppage for necessary maintenance. Some of the nonoperating time was perhaps forced on the boiler operator; for example, because operating was not profitable. Unplanned stoppage is often the major culprit in unavailability.

The time we tally to availability calculation is the time during which the boiler could have operated. This means in addition to the time the boiler produced steam, the time it could have produced steam but did not need to. We define forced outage as nonproduction time caused by unforeseen occurrences. Unforeseen occurrences are most often some type of equipment deficiency.

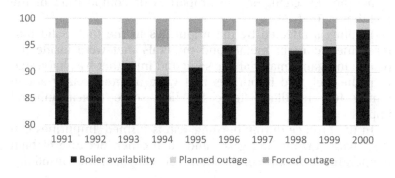

Figure 8.1 Availability of a boiler. After Maryamchik and Wietzke (2000).

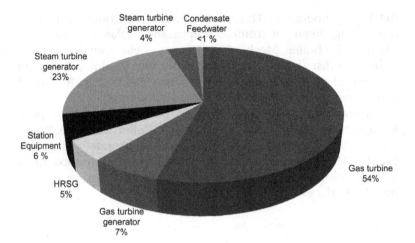

Figure 8.2 An example of major equipment's contribution to unavailability in gas turbine combined cycle cogeneration plants.

In Fig. 8.2, the unavailability of gas turbine combined cycle plants is shown. As can be seen, the rotating equipment is the major source of unavailability. It is, however, worth noting that the major concern is not scheduled maintenance but unscheduled outages.

8.2 Heat Transfer Surface Fouling

Fouling means the reduction of heat transfer when tubes get covered with ash deposit. Fouling reduces heat transfer because the deposit functions in a similar way to adding insulation. If the deposition gets large, then the flue gas velocity and pressure drop increase, and in extreme cases the flue gas passage might be blocked. Sometimes extensive fouling can damage individual tubes or heat transfer surfaces because of uneven thermal expansion, heat flow, and rough removal methods.

Fouling is prevented by proper sootblowing, Fig. 8.3, and sometimes by changing the properties or composition of the fuel that is fired.

Fouling is affected by the impurities in the fuel. Solid biomass fuels contain metals and minerals that were inside the plants, inorganic material like sand and mud that was harvested with the fuel, and impurities that were unintentionally added during fuel handling. Table 8.1 lists some typical mineral impurities.

In the furnace during burning, ash is formed. Impurities can react with surrounding gases and each other. Because of high temperatures during combustion, ash can melt and resolidify.

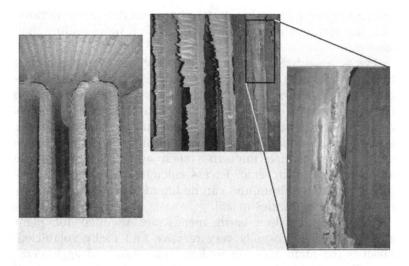

Figure 8.3 Fouled economizer surface.

Table 8.1 Ultimate and Proximate Analyses for Different Types of Fuel

Mineral	Sludge	Wood	Grass	Straw	Husks	Bark	Peat	Coal
Moisture, wt%	64.3	15.9	30.7	10.4	9.0	13.0	41.2	8.1
Ash, wt% dry	19.4	0.6	3.6	8.6	10.8	3.7	5.5	11.8
Volatile, wt% daf	87.8	84.1	83.5	81.1	79.8	76.5	73.9	38.4
HHV, MJ/kg daf	18.8	20.3	19.7	19.5	20.5	21.3	22.9	33.8
LHV, MJ/kg daf	17.9	19.0	18.5	18.1	19.1	19.7	21.4	31.1
C, wt% daf	51.2	51.2	49.6	48.8	50.4	53.8	57.4	79.4
H, wt% daf	6.24	6.15	5.72	5.99	6.28	5.84	6	5.29
O, wt% daf	41.4	42.4	43.9	43.9	42.6	40	35.5	12.2
N, wt% daf	<1.5	<0.5	<1.5	<1.5	<1.5	<0.5	1.9	1.5
S, wt% daf	<1.5	<0.5	<0.5	<0.5	<0.5	<0.5	0.3	1.4
Cl, wt% daf	0.43	0.027	0.196	0.496	0.143	0.022	0.059	0.25
Si, mg/kg dry	46,000	N.A.	6775	17,025	14,000	422	12,615	25,148
Al, mg/kg dry	27,700	N.A.	100	1579	2700	188	4181	13,123
Fe, mg/kg dry	1747	N.A.	109	1417	2300	90	6387	7255
Ca, mg/kg dry	88,600	N.A.	1273	4694	13,000	13,622	6200	5421
Mg, mg/kg dry	2870	N.A.	534	1818	5100	728	634	1666
Na, mg/kg dry	1725	30	319	610	1090	40	144	1142
K, mg/kg dry	1652	680	7633	11,634	22,233	1627	548	1287

Source: From Theis, M., 2006. Interaction of biomass fly ash with different fouling tendencies. Ph.D. Thesis, Åbo Akademi, Report 06—02. ISBN 9521217308 (Theis, 2006).

Ash from biomass burning is often hard and difficult to remove. Ash that contains a molten fraction is often more fouling than solid ash.

Plants absorb SO_2 from the atmosphere. Water absorbed through roots contains minerals such as sulfates. However, neither route leads to high sulfur levels for biofuels. Chlorine in biofuels similarly enters with water as a chloride ion and its concentration depends on the nutrient composition of the soil. Phosphorus is typically utilized as a fertilizer. Silicon in plants enters either though impurities such as soil or though the absorption of silicic acid. Excess calcium can accumulate as oxide or oxalate. Aluminum can be found only in small quantities from the impurities in soil.

Alkali and alkaline earth metals are required for plant growth. They are usually very reactive and easily volatilized. Alkali is the main agent in fouling, corrosion, and agglomeration. The potassium content of agricultural biomasses such as grasses and straw is high.

8.3 Gas Side Corrosion of Heat Transfer Surfaces

Typically the aim of corrosion studies is either to find a material that can withstand the existing corrosive environment or to change operation so that the corrosive environment is eliminated. The main problem in finding what is actually occurring is the limited real-time access in an operating boiler to observe the corrosive environment. Therefore laboratory studies are used to find the cause and to compare various materials (Skrifars et al., 1998). Often only experience gives the justification for choosing a particular material (Sonnenberg, 1968).

8.3.1 Normal Tube Surface

As corrosion is defined by the destruction of a normal tube surface, we should look at its appearance in a typical boiler, Fig. 8.4. The outermost surface of a boiler tube oxidizes fast. Generally a somewhat protective layer, which prevents oxygen reaching the unreacted tube, is formed. Protection from oxygen diffusion is higher if the Cr, Ni, and Mn content of the tube is increased.

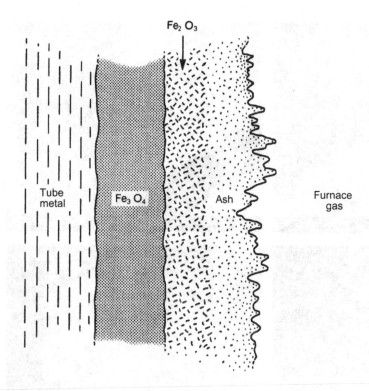

$Fe_2 O_3$

Tube metal

$Fe_3 O_4$

Ash

Furnace gas

Figure 8.4 Typical tube surface towards the furnace.
After Akturk et al. (1991).

8.3.2 High-Temperature Corrosion

High-temperature corrosion in the gas side occurs particularly in the superheaters and in the furnace but also elsewhere when the flue gas temperature is higher than about 500°C. High-temperature corrosion is often caused by a combination of problematic ash deposits and corrosive gaseous compounds. High-temperature corrosion can be caused, e.g., by chlorine (Cl) which reacts easily to corrosive gases such as hydrochloride (HCl) (Bruno, 1999). Alkali metals (Na, K) form deposits that can have a molten phase close to 500°C. For the operator the main problem is that some biofuels can contain problematic minor impurities. As the overall ash content is low, these impurities can enrich to ash deposits. A typical example is biosludge that has caused severe superheater corrosion even though the share of the biosludge in the total fuel was only a few percent (Salmenoja and Mäkelä, 1999).

Deposits are not uniform, but consist of a collection of materials that arrive at the surface by various means. The hot deposits themselves can react with flue gas compounds (Sharp et al., 2012) (Fig. 8.5).

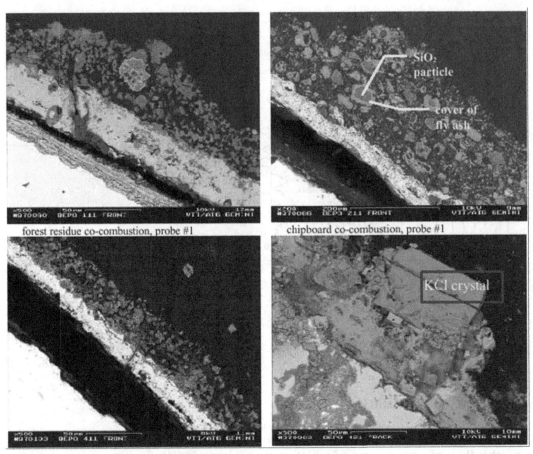

Figure 8.5 Ash deposits on a superheater showing deposits at various magnifications and fuels; upper left forest residue with chips, upper right chipboard with chips, lower left forest residue with chips when low deposition, lower right chipboard with chips high magnification.

Engineering experience has been to select higher chromium-containing alloys for places that suffer from corrosion. As chrome (Cr) and nickel (Ni) cost significantly more than iron, this increases costs significantly. The choice of suitable material is still a matter of trial and error. The main problem in corrosion science is that gas and deposit sampling during the corrosive event is often difficult and seldom even attempted. Therefore actual data on operating boilers is scarce.

8.3.3 Acid Dew Point

Low-temperature gas side corrosion is usually found in the economizers and air heaters but it also occurs in other surfaces where the flue gas temperature is lower than about 200°C. Low-temperature corrosion is often associated with the formation of acidic deposits. Acid dew point means one can measure low pH on the tube surface. Typically acidity can be seen in the economizer, air preheater, or flue gas duct areas where the temperature is around 100°C. When flue gases cool, then gaseous species start condensing. Typically the sulfur dew point refers to conditions where sulfur trioxide starts condensing. Also HCl condensation at the same temperature regime is possible.

8.3.4   Alkali Corrosion

Particularly in secondary and tertiary superheaters, sodium and potassium form ash compounds that have first melting points in the range of 500–600°C. If we have chloride, then the formed compounds can have very low ash-melting points (Rönnquist, 2000). The actual corrosion can start when the temperature is close to the first melting point of one of the ash species in the deposits. As the temperature increases, then typically more and more ash is in molten form (Skrifars et al., 2008). When the liquid content of the ash gets higher than about 70%, it flows down along the superheater surface. Molten ash corrosion shows as uniform thinning and is often caused by the high potassium content in biofuels (Orjala et al., 2001).

This corrosion is not seen at the windward side of the tube because there the deposit is thick. A thick deposit has a low heat flux and molten ash cannot penetrate the tube surface before it cools and solidifies. In alkali corrosion the corrosion rate depends on the flue gas composition, the tube surface temperature, and especially on the local ash properties. The corrosion rate between tubes and along a single tube can therefore vary widely.

8.3.5   Chloride Corrosion

Biomasses often contain chloride as salts readily enter growing plants along with water. During burning in the furnace the chlorine reacts and forms gaseous Cl_2 and HCl. Iron in the tube surface can react with these gases to form ferrochloride (Bankiewicz et al., 2012). Ferrochlorides are stable only in reducing conditions. When oxygen is present, they react back to ferrous oxide, releasing chloride gases. Chloride corrosion has been studied a lot in connection with waste-burning and recovery boilers (Born, 2005).

8.4 Water Side Corrosion and Problems

The boiler steam–water system can experience several types of corrosion. Steam–water corrosion can result in severe damage that will affect the safety and operational reliability of the boiler. In some types the rapid failure of a tube can occur with relatively small overall metal loss (Littler et al., 1992).

Oxygen corrosion causes localized pitting of the tube's inside surface. Even small amounts of dissolved oxygen can cause severe damage. Acid corrosion occurs when there is low pH in the water. This corrosion is characterized by a general wastage of the metal. Caustic corrosion is characterized by high pH in the water. This corrosion is seen as loss of magnetite film and irregular patterns, and is often referred to as caustic gouging. Hydrogen attack results in a loss of metal strength. There are no visible signs to indicate this is happening but often hardness is increasing. All these three corrosion types can be heavily localized and might be assisted by deposits inside the tubes.

8.4.1 Oxygen Corrosion

Oxygen corrosion causes pitting of the tube's inside surface. Even small amounts of dissolved oxygen can cause severe damage. Oxygen is extremely corrosive when present in hot water. Components that raise water temperature rapidly, such as closed heaters and economizers, are particularly susceptible to an oxygen corrosion attack. The temperature increase provides the driving force that accelerates the corrosive reaction.

Out-of-service boilers are also susceptible to an oxygen corrosion attack. Such an attack may typically be found at the water–air interface. If the boiler is shut down for longer periods, special procedures are necessary to avoid corrosion. Complete removal of dissolved oxygen is necessary. Oxygen is removed mechanically in a deaerator and chemically by a suitable chemical dosage.

8.4.2 Acid Corrosion

When low pH water attacks steel, the attack is characterized by a general wastage of the metal, as opposed to the localized pitting nature of an oxygen attack (Vänskä, 2010). Since acidic conditions during operation will produce rapid and severe corrosion, pH control of the boiler water is critical. In the normal operating pH range of boiler water very little corrosion occurs. Under these conditions, steel readily reacts with water to form magnetite:

$$3Fe + 4\,H_2O => Fe_3O_4 + 4H_2$$

Magnetite forms as a thin, tenacious film on boiler steel. This film protects boiler steel against attacks. However, magnetite is no match for the acidic or caustic conditions that can (and do) arise in a boiler, as these conditions readily dissolve the thin, protective film. If the pH in the boiler water decreases below 4.5 an electrostatic precipitator is required as the separated magnetite layer may block the tubes.

In high-purity feedwater systems (where demineralized makeup is used), the unbuffered water is subject to extreme pH swings from minor changes in acid or alkali concentrations. A typical primary constituent of chemical internal treatment is phosphate. Phosphate will serve to buffer the deionized water to dampen pH swings. To control acid corrosion in high-pressure boilers, the pH will be maintained above 8.5. One typical method of control is to feed caustic (NaOH) into the feedwater.

8.4.3 Caustic Corrosion

High-pH boiler water can be as equally corrosive as acid. High-pressure boilers are more susceptible to caustic attack than acidic attack. Deionized feedwater provides no buffering capacity to react with the OH-ion and inhibit caustic concentration. The predominate contaminate in demineralized makeup is sodium hydroxide (caustic). The caustic will become concentrated under scale where evaporation of water occurs. As steam contains almost no caustic the caustic concentration is increased. The concentrated caustic readily "dissolves" the protective magnetite film, forming complex caustic-ferritic compounds. The exposed steel then reacts with water to reform the magnetite film, which is again dissolved by the caustic. The attack on the boiler steel continues for as long as the concentrated caustic exists.

The resultant metal loss assumes irregular patterns and is often referred to as caustic gouging. When deposits are removed from the tube surface during examination, the characteristic gouges are evidence of caustic corrosion. The gouges are accompanied by a typical white salt deposit, which is a combination of sodium carbonate (which is the residue of the caustic after contact with carbon dioxide in the air) and crystallized sodium hydroxide. Prevention of caustic corrosion is done by maintaining correct pH and controlling the molar ratio of sodium in the boiler water.

8.5 Erosion

Erosion means that material on a surface is removed when hit by particles. Sandblasting is a common example of using erosion to remove material from surfaces. The erosion rate increases when the particle velocity increases. It needs a certain kinetic energy to happen. In boilers the erosion depends on the local velocity field and ash characteristics. In addition the ash and solid fuel transport and handling equipment suffer from erosion. In fluidized beds the lower furnace must be protected from erosion. Erosion can be controlled by applying a protective harder surface through welding on the tubes and using refractory-coated surfaces (Fig. 8.6).

Tube panels can be covered with refractory. This has proven to be a good form of erosion protection, especially in a circulating fluidized bed furnace. Refractory also provides a fair corrosion protection (for recovery and grate). It should be remembered that refractory needs regular maintenance.

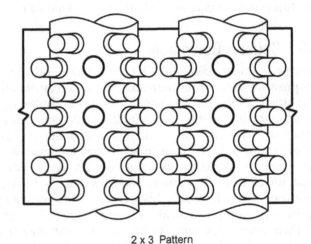

2 x 3 Pattern

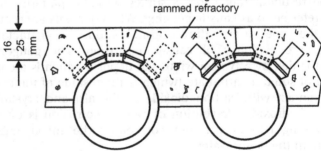

Low cement high alumina rammed refractory

16
25 mm

Figure 8.6 Refractory with studs to protect the furnace wall.

Table 8.2 Hardness Properties of Minerals

Constituent	Mohs Hardness Number	Vickers Hardness (kg/mm^2)
Quartz	7	1200–1300
Silicates		
Kaolin	2–2.5	30–40
Illite	2–2.5	20–35
Muscovite	2–2.5	40–80
Orthoclase	6	700–800
Kyanite	4–7	500–2150
Topaz	8	1500–1700
Alumina	9	2100
Carbonates		
Calcite	3	130–170
Magnesite	4	370–520
Siderite	4	370–430
Pyrites	6–7	1100–1300

The most typical refractory materials are alumina and silica. To improve mechanical and thermal properties, e.g., magnesia can be used. The harder the particle, the higher is the erosion rate. In Table 8.2 the hardness properties of typical minerals are presented. The hardest substances are alumina, quartz, and pyrite. The choice of refractory is based primarily on ease of installation and lifetime at the intended location. Often the time needed to replace refractory is only a few days, so materials that need days to dry are usually not considered.

8.6 Corrosion Prevention and Surface Examination

Typically we can predict corrosion by regularly inspecting surfaces known to have problems. This involves short time interval checks with the results tabulated. We select the locations based on boilers that are similar or that operate in a similar environment, and try new materials. Often it is known from experience that a particular surface in a certain type of boiler is prone to this type of corrosion. If corrosion is observed, then one can start thinking about how to prevent the corrosion or at least lower the corrosion rate. The main thing in corrosion prevention is learning from experience. When heat transfer surfaces are inspected, their condition can be ascertained. Usually inspection shows corrosion phenomena occurring in certain places. Appropriate action can then be taken before anything serious happens.

Surface assessment can be done during maintenance stops. The most important thing is to keep a log of measurements and findings. Typically nondestructive testing (NDT) methods used are ultrasonic inspection (tube thickness measurement), radiographic examination (welds), and surface examination with penetration fluid (cracking). Not every weld is inspected. Usually codes allow higher allowable stresses if all welds are inspected. This must then be weighed against the cost of the inspection.

In modern boiler manufacture and maintenance, quality control is essential. Many firms certify quality control systems. One needs to make sure that the people doing the work are knowledgeable and properly certified. They need to follow accepted procedures and all work must be recorded (Fig. 8.7).

Figure 8.7 Feedwater treatment plant.

8.7 Feedwater Treatment

Boiler water treatment is done to minimize corrosion, deposition, scale, and carryover in the water–steam circuits. The aim is to maximize the reliability of the steam–water circuit.

Boiler water treatment has basically two phases: First, we remove impurities from the water to the desired cleanliness. Second, we add chemicals to the feedwater to adjust the pH, create less favorable conditions for deposition, and remove oxygen. Special water treatment chemicals are used in almost all boilers. Lastly to lower further the mineral content demineralization can be applied. Water treatment involves a lot of small subprocesses and equipment:

- Aeration: Mixing air into water oxidizes dissolved salts and makes them filterable. Aeration also removes undesirable gases as carbon dioxide.
- Coagulation: Adding chemical coagulating materials reduces surface-water contamination in the form of coarse suspended solids, silt, turbidity, color, and colloids.
- Filtration: Filters separate coarse suspended matter from raw water.
- Chemical softening: Chemicals can remove hardness, silica, and silt from makeup water.
- Demineralization: cation and anion exchanges are used to remove ionized mineral salts.

8.7.1 Aeration

Dissolved salts and gases in the raw water can be removed by aeration. Mixing water and air oxidizes these salts and makes them filterable. Carbon dioxide is also removed during aeration. The solubility of gas in water is directly proportional to its partial pressure in the surrounding atmosphere. The partial pressure of a gas such as carbon dioxide is low in a normal atmosphere. Establishing equilibrium between water and air by aeration results in the saturation of water with oxygen and nitrogen and results in the practical elimination of gases such as carbon dioxide and hydrogen sulfide. Increasing the temperature, the aeration time, and the surface area of the water helps with the removal of gases.

8.7.2 Coagulation

Adding chemical coagulating materials reduces surface-water contamination in the form of coarse suspended solids, silt, turbidity, color, and colloids. The chemicals form a floc, which assists in agglomerating impurities. Settlement of the particles permits a clear effluent from the coagulating chamber. Some chemicals used for coagulation are filter alum, sodium aluminate, ferrosul, activated silica, and various proprietary organic compounds. Temperature, pH, and mixing affect the efficiency of coagulation.

8.7.3 Filtration

Filters separate coarse suspended matter from raw water, or remove floc or sludge components from coagulation or process softening systems. Generally, gravity and pressure-type filters are used for this purpose. Beds of graded gravel or coarse anthracite are the common materials used in the filter bed. Diatomaceous earth and special precoat filters are generally used to remove oil and reduce color in feedwater makeup.

8.7.4 Chemical Softening

Nonscaling feedwater can be obtained with proper pretreatment of raw water. In chemical softening, various chemical combinations are added to raw water to remove hardness, silica, and silt from the makeup water. Economics and boiler operating conditions dictate the technique selected for the application. Additives can contain, e.g., lime, phosphate, or zeolite.

8.7.5 Demineralization

In demineralization, ion exchange removes ionized mineral salts. Cations as calcium, magnesium, and sodium are removed in the hydrogen cation exchanger; and anions as bicarbonates, sulfates, chloride, and soluble silica are removed in the anion exchanger. Synthetic cation and anion exchange resins are used in the demineralization of water. The cation exchanger is regenerated with acid, while the anion exchange material is regenerated with caustic. Demineralization can yield a pure water, equal or superior to the best evaporated water. The anion and cation resins can be arranged in various combinations to produce the best water most economically. Two-, three-, or four-bed units, or a single mixed-bed demineralizer can be used to accomplish the required result (Fig. 8.8).

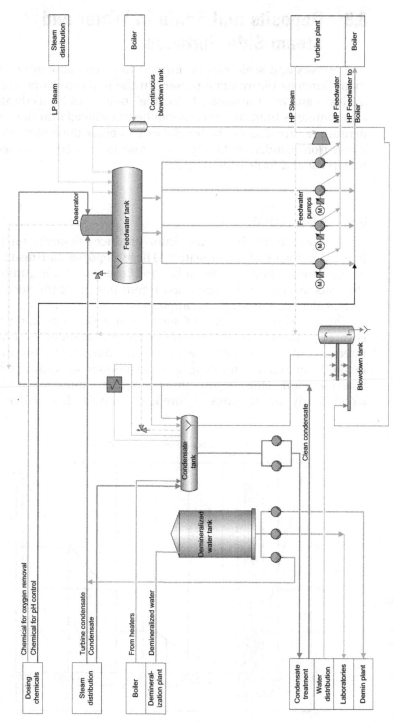

Figure 8.8 Typical boiler feedwater system.

8.8 Deposits and Scale in Water and Steam Side Surfaces

Deposits and scale can be formed from contaminating elements such as bicarbonate present in the makeup water. Metal oxides can be transported to the boiler with feedwater. Contaminants from the process can be introduced into the condensate returned to the boiler. Solids can enter the boiler circulation from condenser leakage. All these foreign elements tend to deposit and form scale (Fig. 8.9).

8.8.1 Carryover

Carryover from the steam drum produces deposits on the internal surfaces of superheaters. These deposits can result in severe overheating of the tube and even in tube rupture. Deposits are caused by entrained impurities in steam, such as silica, calcium, and sodium compounds.

The mechanical aspects of steam and water separation are covered in the section 4.4 describing the operation of the drum internals. This section deals with the phenomenon of carryover, methods of steam sampling, and techniques of steam-purity determination. Once again, these subjects cannot be understood solely by rigorous theoretical analysis. Knowledge of

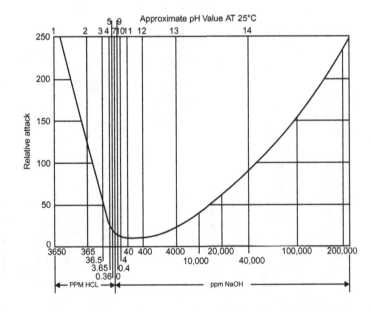

Figure 8.9 Rate of corrosion of carbon steel in boiler water as a function of pH. After Navitsky, G.J., Gabrieli, F., 1980. Boiler water treatment, feedwater treatment, and chemical cleaning of drum-type utility steam generators. Combustion, August; p. 19 (Navitsky and Gabrieli, 1980).

laboratory testing procedures and the ability to interpret field operating experience are required.

Carryover of boiler water in steam leaving the drum provides a path for introducing solid materials into the superheater section. Modern separator designs can mechanically reduce the moisture content of the steam to 0.2% or less. However, in addition to the mechanical carryover of boiler water, another mechanism exists, which results in the contamination of steam with solid materials.

Priming is the development of excessive moisture in the steam because of spouting or surging of the boiler water into the steam outlet. This is a rare, easily identified type of carryover. It is usually promoted by too high a water level in the drum, spouting of a submerged riser, or a sudden swelling of the water in the boiler after a drop in pressure or a sudden increase in rating. Priming is rarely, if ever, associated with boiler-water concentration.

Spray carryover, mist, or fog are degrees of atomization of the boiler water. Mist is carried from the drum by the steam like dust is carried by air currents. This carryover is present to a degree in all boilers, and it is the function of the drum internals to separate and filter out such spray before the steam leaves the drum. Development of spray carryover indicates failure of the drum internals due to exceeding the velocity limitations of the purification equipment. It is characterized by initial development below the full rating of the boiler and it continues to increase with boiler load. Spray carryover is not sensitive to boiler-water concentration below the foaming range. Improved drum internals are capable of reducing the steam-borne mist to a value as low as a few parts per billion of solids.

Leakage is a general term applied to the bypassing of impure steam or boiler water through the drum internals. Normally localized, this form of carryover is directly related to poor design or installation of the drum internals. At times the local contamination may not be sufficient to be reflected in steam-purity measurements of total steam flow. A careful inspection of the drum internals will usually reveal this source of carryover. Where the leakage is sufficient to register impurity tests of steam, it will be found that the impurity increases slowly with rating and is relatively insensitive to changes in water level and boiler-water concentration.

As operating pressures increase, the steam phase exhibits greater solvent capabilities for the salts that may be present in the water phase. These salts will be partitioned in equilibrium between the steam and water, a phenomenon known as vaporous carryover. Vaporous carryover will contribute an additional quantity of boiler water solids directly to the steam, independent of the efficiency of the steam–water separation components.

Silica was the first material found to exhibit significant vaporous carryover. Silica fouling of turbines was common until it was recognized that successful control of the amount of silica in the steam could only be accomplished by controlling the amount of silica in the boiler water. A similar solution will be required for other solids when operating at high pressures.

8.8.2 Foaming

Foaming is the development of excessive moisture in the steam from foam from the drum. It is the main form of carryover in low-pressure units, in which the boiler water may contain high concentrations of dissolved solids, and is the most troublesome and most erratic type. Foam forms in the steam-generating sections of the boiler when the water-films around the generated steam bubbles are stabilized by the impurities in the boiler water. Boiler circulation carries this foam up to the boiler drum, where it tends to accumulate at the water level. The foam produced may entirely fill the steam space of the boiler drum or it may be of a relatively minor depth.

The bulk water in the circulating mixture entering the drum is readily separated, but the wet emulsion of very small foam bubbles collects at the water level to a depth largely dependent on the rate of drainage of excess water out of the foamy mass. A considerable amount of moisture is trapped in the foam. When foam carryover occurs, it is frequently sudden and excessive, and the steam sample registers a solids content characteristic of boiler water.

In general, foam carryover from a boiler can be avoided by keeping boiler-water concentrations within the range suggested by the boiler manufacturer. These specifications, of course, cannot be a guarantee against foaming, which, as indicated previously, is primarily a chemical problem.

8.8.3 Identification of Carryover Type

A systematic field investigation can identify carryover. Steam flow, water level, and boiler-water concentrations are the three major factors that can create carryover. By varying these three factors, one at a time, test results can usually be interpreted to determine the specific source of the carryover condition.

Steam flow establishes the velocity distribution in the boiler drum. Excessive steam flow can increase steam velocity to a point where entrained moisture can overload the dryer. High water can create spouting and excessive carryover. This can occur at low steaming rates and boiler-water concentrations.

Foaming is a characteristic of boiler-water concentration. With the water level and steaming rate at the recommended values, any carryover that can be precipitated or eliminated by a change in boiler-water concentration can be attributed to foaming.

8.9 Managing Steam and Water Quality

8.9.1 Feedwater and Boiler Water Quality Control

The service life and availability of a steam power plant depend on the quality of feedwater, and thus careful management of water treatment is essential. Treating equipment has to be operated and maintained according to the quality requirements of the feedwater and raw water available.

Treatment of feedwater is used to provide good operating conditions for the water–steam turbine and to produce suitable water or steam for utilization. This is accomplished by preventing corrosion of the boiler and other water–steam circuits, accumulation of deposits, and foaming of boiler water. Magnetite film formed in the insides of the tubes in the boiler water–steam circuit protects it from corrosion. The conditions have to be such that an unbroken magnetite film is maintained.

Boiler water quality is controlled by feedwater quality and boiler blowdown. Practically all impurities entering the boiler with the feedwater stay in the boiler and decrease the boiler water quality. Boiler water quality can be improved by blowdown where a small saturated water flow is extracted, typically from the drum. Boiler water should not exceed the specified maximum values for the boiler. Typically the impurity content in boiler water correlates inversely to the blowdown and directly with the impurity content in the feedwater.

One of the most corrosive agents is O_2 in the feedwater. This will cause corrosion of the feedwater pipes and economizer tubes. If high O_2 in the feedwater is not caused by improper sampling, then it is often caused by deaerator malfunction. This can be caused by too small a vent steam flow or too low a steam flow to the deaerator. The deaerator internals may also be in poor condition. We can also control the O_2 in the feedwater entering the boiler by utilizing some suitable chemical as an additive.

The most important value to watch is the pH value of the feedwater. The pH value needs to be maintained at the proper window to prevent the destruction of the magnetite film and corrosion. The pH can be maintained by utilizing the right amount of chemicals and proper blowdown. If the feedwater pH goes below 8.0–8.5, then drastic problems in the boiler can occur.

In addition to the pH value we must continuously monitor the conductivity of the feedwater. If conductivity in feedwater increases suddenly, in general raw water is leaking into the condensate system or the water treatment system is malfunctioning. One must check the feedwater hardness and silicate content immediately and take necessary measures accordingly.

8.9.2 Steam Quality Control

Steam contains by nature fewer contaminants than water. However, high-pressure steam is susceptible to containing some silica, sodium, and copper. These might cause internal deposits, result in corrosion of superheater tubes, or even get carried all the way to the turbine, where they deposit on the turbine blades. Steam quality guidelines can be found from standards and from vendor data.

If there is too high SiO_2, the consequences are deposits on the turbine blades. Deposits on turbine blades cause excess vibration and require turbine shutdown for cleaning and maintenance. The reason can be too high a SiO_2 content in the boiler feedwater. This can be corrected by increasing boiler water blowdown, improving feedwater quality, or reducing the boiler water pressure to correspond to the SiO_2 standard; in an extreme case one can use a turbine bypass. The reason can also be lack of superheated steam temperature control or poor spray water quality.

If the amount of nonvolatile salts is too high, then they may deposit on the superheater and turbine blades. The reason can be foaming in the boiler drum. This can be controlled by increasing blowdown or adjusting the drum water level, which might mean boiler load reduction. Another reason can be the poor condition or inadequacy of droplet separators and demisters.

8.9.3 Condensate Quality Control

Return condensate is derived from supplied process steam. In practice there is always a danger of something leaking to lower that atmospheric part of the condensate system and a danger of inadvertent pumping of dirty water to the condensate system. Because of the danger of unwanted process condensates entering the condensate system, the use of both pH and conductivity measurements to monitor condensate quality is recommended. If conductivity values are exceptionally high, take the measures specified for feedwater. Control the salt-, Fe-, and Cu-content of the various condensates with regular analyses and, if the values are high, take the measures described in the section 8.7 on feedwater. If the oxygen content increases, make sure that proper oxygen sampling takes place and that no oxygen is absorbed into the sample water at this stage. If the oxygen content in the condensate is exceptionally high, increase the dosage for the oxygen scavenging chemical and check the operation of the deaerator.

References

Akturk, N. U., Allan, R. E., Barrett, A.A., Brooks, W. J. D., Cooper, J. R. P., Harris, C. P. and Willis G., Modern power station practice. vol. B, Boilers and Ancillary Plant, third ed. Pergamon press, Singapore, 184 p. ISBN 0080405126.

Bankiewicz, D., Vainikka, P., Lindberg, D., Frantsi, A., Silvennoinen, J., Yrjäs, P., et al., 2012. High temperature corrosion of boiler waterwalls induced by chlorides and bromides—Part 2: lab-scale corrosion tests and thermodynamic equilibrium modeling of ash and gaseous species. Fuel. 94, 240–250.

Born, M., 2005. Cause and risk evaluation for high-temperature chlorine corrosion. VGB Power Tech. 85 (5), 107–111.

Bruno, F., 1999. On the influence of chlorides and sulphurous compounds on the corrosion of superheater tubes in boilers with special consideration on kraft recovery boilers. Technical Report, VF-664, Värmeforsk Service AB. 36 p.

Littler, D.J., et al., 1992. Modern power station practice : chemistry and metallurgy, third ed. Incorporating Modern Power System Practice, vol. E. Pergamon Press, Inc, Oxford England, 576 p. ISBN 0080405150.

Maryamchik, M. and Wietzke, D.L., 2000, Circulating fluidized bed design approach comparison. POWER-GEN International 2000, November 14-16, 2000, Orlando, Florida, U.S.A. 6 p.

Navitsky, G.J., Gabrieli, F., 1980. Boiler water treatment, feedwater treatment, and chemical cleaning of drum-type utility steam generators. Combustion. 52 (2), 19.

Orjala, M., Ingalsuo, R., Paakkinen, K., Hämäläinen, J., Mäkipää, M., Oksa, M., et al., 2001. How to control superheater tube corrosion in FB boilers which use wood and wood waste as fuel. 10th International Symposium on Corrosion in the Pulp and Paper Industry, VTT Symposium Series, Otamedia Oy, Espoo 2001. ISBN 9513857204.

Parrot, G.F., 1829. Mémoire concernant de nouveaux moyens de prévenir tous les accidens qui ont lieu dans les machines à vapeur, et nommément sur les pyroscaphes, causés par un excès d'élasticité des vapeurs. Ph.D. thesis, St. Pétersbourg, Imprimerie de l'Académie des Sciences, 20 p.

Rönnquist, E.M., 2000. Superheater corrosion in biomass boiler—theories and tests in Västermalmsverket, Falun (Överhettarkorrosion i bioeldad panna—teorier och prov i Västermalmsverket, Falun). Technical report, SVF-708, Värmeforsk Service AB, 94 p (in Swedish).

Salmenoja, K., Mäkelä, K., 1999. Corrosion in fluidized bed boilers burning potassium and chlorine containing fuels. TAPPI J. 82 (6), 61–166.

Sharp, S., Singbeil, D.L., Keiser, J.R., 2012. Superheater corrosion produced by biomass fuels. NACE International COROSION/2012 conference, Salt Lake City, UT, March 2012, 18 p.

Skrifars, B.J., Backman, R., Hupa, M., Sfiris, G., Åbyhammar, T., Lyngfelt, A., 1998. Ash behaviour in a CFB boiler during combustion of coal, peat or wood. Fuel. 77 (1/2), 65–70.

Skrifars, B.J., Backman, R., Hupa, M., Salmenoja, K., Vakkilainen, E., 2008. Corrosion of superheater steel materials under alkali salt deposits. Part 1: The effect of salt deposit composition and temperature. Corros. Sci. 50 (5), 1274–1282.

Sonnenberg, G.S., 1968. Hundert jahre sicherheit (Hundred years of safety), Technikgeschichte in Einzeldarstellungen, 6. VDI-Verlag, Düsseldorf, 338 p. (in German).

Theis, M., 2006. Interaction of biomass fly ash with different fouling tendencies. Ph. D. Thesis, Åbo Akademi, Report 06-02. ISBN 9521217308.

Vänskä, J., 2010. Voimalaitoksen palamisilman esilämmittimen sisäpuolisen korroosion hallinta pinta-aktiivisten amiinien avulla (Internal corrosion control of steam heated combustion air pre-heater in power plant with surface active amines). M.Sc. Thesis, Lappeenranta University of Technology, Energy Technology, Lappeenranta, 120 p. (in Finnish).

9

DIRECT AND GRATE FIRING OF BIOMASS

If the biomass is shredded to small enough particles, it can be burned in a flame. Direct firing of biomass has been tested at large coal-fired units as a cheap replacement for coal in case fossil carbon dioxide emissions from coal become unacceptable. Direct firing of biomass pellets is used in various boiler designs up to 60 MWth. As the biomass market is changing towards standardized fuels and pellets become more widely traded around the globe, the direct firing of biomass will increase in popularity.

Grate firing is the oldest type of firing. Grate firing was the main combustion technique up until the 1930s when the pulverized firing of coal (PFC) replaced it in utility boilers. For industrial boilers burning solid biomass fuels, grate firing continued as the main combustion method, although in large units it has since the late 1980s' been replaced by fluidized bed firing. Grate firing of solid biomass fuels is still the most popular boiler type for smaller boilers up to about 10 MWth.

In grate or stoker-fired boilers, the combustion of solid fuel occurs in a bed at the bottom of the furnace. Over the centuries, numerous different technical designs for grate firing systems have been invented for burning various kinds of solid fuel (Ostendorf, 1986). In all designs, the fuel burns in a layer, at the bottom of an enclosure, through which some of the air for combustion passes.

Steam Generation from Biomass. DOI: http://dx.doi.org/10.1016/B978-0-12-804389-9.00009-5

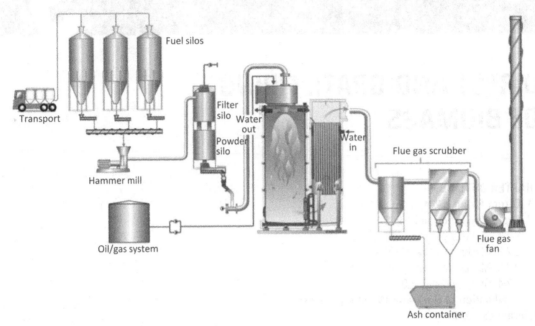

Figure 9.1 Direct firing of pellets. Courtesy of Valmet Power.

The benefit of grate firing is that all forms of solid fuel can be cheaply fired. Even low-grade fuels such as peat and bark can be fired if their properties remain somewhat constant. The main disadvantage of grate firing is the slow change in firing rate and the low burning rate in the grate requiring a large grate area (Päällysaho, 2009). As all combustion processes occur sequentially, there is always a relatively large amount of unburnt fuel in the grate. Changing the bed burning rate is therefore challenging (Fig. 9.1).

9.1 Direct Firing

Pellets can be directly fired in a burner resembling pulverized coal fired (PFC) boiler burners. The pellets are first broken apart at a hammer mill, transported to a silo, and then blown into a burner to be burned in a flame inside the furnace. As with many other solid biomass boilers, the startup needs liquid or gaseous fuel. In this case it is oil or bio-oil.

The startup and the load control are very fast and the dry fuel allows a hot flame with highly efficient combustion. As pellets are readily available from various commercial sources, direct firing can be remotely operated, and the investment cost is low, these boilers have started to gain acceptance as peak load and backup plants for larger combined heat and power (CHP) boilers.

9.2 Grate Constructions

The mechanical construction of the grate can be stationary or moving. Stationary grates, such as the inclined grate, Fig. 9.2, are mainly utilized in small boilers. Mechanical grates are the dominant type of grate for steam generation (Raiko et al., 1995).

9.2.1 Stationary Grate

The stationary grate was the first grate type to be used when steam generation started more than 200 years ago. They are easy to construct but require constant attention and stoking to keep burning stable. Stationary grates employ gravity to move fuel. For solid biomass fuels such as bark and wood chips this requires 30–50 degrees inclination from horizontal Huhtinen (2000). The inclination of the grate depends on the fuel and its ability to flow during combustion. The inclination can change at different locations of the grate. It is typically higher at the fuel receiving end of the grate. To complete the burning of fuel, many inclined grates have a small horizontal grate after the inclined section. This section is called the dump grate.

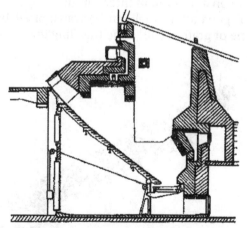

Figure 9.2 Inclined grate, showing fuel feeding up from the left, gravitationally sliding to the right, and ending at a dump grate.

9.2.2 Mechanical Grate

Larger mechanical grates used in steam generation contain moving parts and are equipped with automatic fuel feed and ash removal. The mechanical grate is almost always inclined. In a typical design of a mechanical reciprocating grate, various parts of the grate can be mechanically moved backward and forward to facilitate biomass bed movement on the grate, Fig. 9.2. By changing the speed of the mechanical movement it is possible to regulate the movement of the fuel on the grate. In large grates the speed of the movement can be different in different sections of the grate. A mechanical inclined grate therefore does not need to have as deep an inclining angle as the stationary grate. A suitable angle is 15 degrees Huhtinen (2000). The mechanical grate is one of the most typical grates for biomass firing.

A side view of a large industrial mechanical grate is seen in Fig. 9.3. Fuel feeding is from the top left. The moving grate transports fuel to the lower right. The speed of transport is defined by the operating frequency of the mechanically moving parts. The ash ends at a dump grate, which consist of a perforated cylinder.

A step grate is one example of a mechanical grate. Its name derives from the fact that it looks rather like a large staircase with lots of steps from the bottom to the top. In a step grate each step is made of cast iron grate bars. Air is introduced between the grate plates. The most famous "brand" of mechanical inclined step grates has been the Kablitz grate. The Kablitz grate was the first to make continuous burning of biomass possible without constant human intervention.

Instead of gravitation, fuel can be transported by a moving belt. This type of grate is called the traveling grate. The traveling

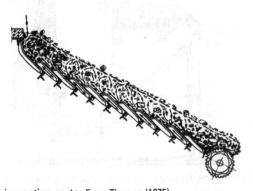

Figure 9.3 Reciprocating grate. From Thomas (1975).

grate has solid elements joined to a chain, which moves horizontally and transports fuel. Fuel is most commonly fed with a spreader stoker on the grate. Changing the rate of fuel addition changes the fuel layer thickness. For coal, a suitable thickness is 10–20 cm, and for wood it is 30–90 cm Huhtinen (2000). The speed of the traveling grate is chosen so that the burning can be completed within the grate.

Another common type of grate is the roll grate. Instead of a stationary or back-and-forth rotating surface the grate consists of giant rolls. These mix the bed efficiently. Even though roll grates are usually built inclined, they can also be built horizontally. Roll grates have found their niche in waste burning as they are insensitive to large objects placed within the waste.

9.2.3 Modern Innovative Grates

A fairy new invention is the horizontally rotating pile grate. In it the fuel is fed from the bottom to form a large pile or hill in a refractory-lined precombustion chamber. Round grate sectors are rotated horizontally. The frequency of back-and-forth movement causes the burning biomass to flow downwards. This type of grate has been successfully used to burn wood and bark. So far this type of grate has proven itself useful for thermal loads of 1–20 MW (Fig. 9.4).

Yet another example of a fairly new type of grate is the vibrating grate (Caillat and Vakkilainen, 2013). A vibrating grate has a back-and-forth movement that pushes fuel particles down a slope. Often the grate is mounted on leaf springs. Grate sections are vibrated by rods or other moving mechanisms. Due to vibration the fuel bed moves from the entry point toward the lower end of the grate. Often the vibrating frequency is adjustable and can be matched to the firing rate and the fuel characteristics.

9.2.4 Grate Cooling

Small grates are uncooled, but all grates for steam generation need cooling. Cooling can be by primary air or by water.

9.3 Combustion of Bark and Wood on a Grate

The burning of a solid biomass fuel on a mechanical grate follows the same stages that can be seen in any combustion.

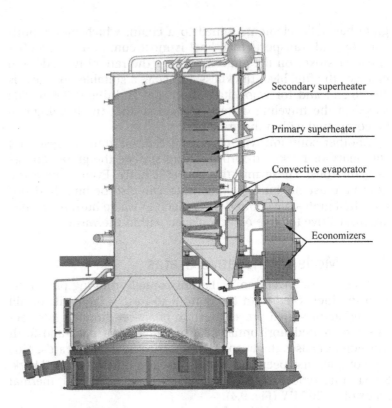

Secondary superheater

Primary superheater

Convective evaporator

Economizers

Figure 9.4 Pile grate. Courtesy of Valmet Power.

In Fig. 9.5, starting from the top left the following stages can be seen: drying of the fuel where the moisture is evaporated, generation of volatile matter leading to a visible flame, combustion of char where the hot char glows with heat, and finally ash reactions (Yin et al., 2008). Typically for solid biofuels one uses refractory, which, when hot, radiates back heat and helps to keep combustion stable and especially allows moist fuels to dry faster. All these phases occur in sequence for a single small fuel particle, although fuel particles are simultaneously at different burning phases on the grate. Large fuel particles can still have fresh fuel in the core while the char on the surface is burning (Horttanainen et al., 2000). The primary air is supplied under the grate. Usually the amount can be controlled in one or more separated zones. The secondary and tertiary airflows are typically injected in the furnace proper above the fuel bed.

In grate firing, irrespective of the type of fuel, combustion takes place at distinct stages, depending on the location of the fuel on the grate (Kortela and Marttinen, 1985). In the fuel entrance region it slowly dries and ignites when the temperature has sufficiently increased. The challenge is to transfer heat

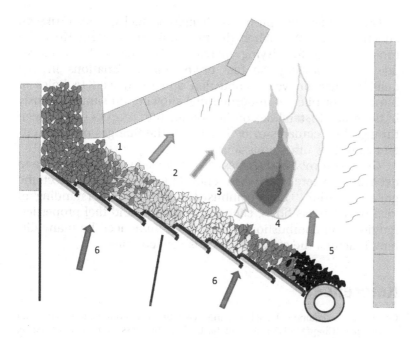

Figure 9.5 Stages of solid biomass combustion in an inclined mechanical grate: 1, fuel feed; 2, drying; 3, devolatilization; 4, char combustion; 5, ash; 6, primary air.

Table 9.1 Achieved Loading Per Grate Area

Fuel	Heat Load per Unit Area (MW/m^2)
Coal (traveling grate)	1.6
Bark and wood (moisture content 60%)	0.4
Bark and wood (moisture content 30%)	0.8

from the burning fuel to drying. Typically this is achieved mainly by the heat radiation. As direct radiation is often not possible, it is typical to position brick walls to reflect heat. Burning in stages means there are large variations in local temperatures and concentrations of flue gases in the furnace. A low heat flux to drying means that the grate area required for moist fuels is significantly higher than for dry files, Table 9.1.

The maximum fuel burning rate depends on emission limits as well as on the heating rate of the biomass. Articulate emissions as unburned carbon increase when the burning rate is increased. This is because air velocity through the grate needs to be higher for fractionally larger fuel particles and still unburned carbon to entrain into flue gases from the burning fuel in the grate.

Grate boilers are prone to changes in fuel quality (Eriksson and Ingman, 2001). When the fuel moisture content rises, the time required for drying increases. The flame front moves downward, slowing the fuel drying more. Variations in fuel quality cause uneven combustion and often lead to holes of burnt ash or piles of unburned fuel showing up simultaneously in the grate. Instabilities of the combustion process can be seen through bed cameras or by looking at the flue gas oxygen content as well as other emissions.

Grate controls, bed shapes, disturbance location, and time to determine where and when unevenness occurs all affect the grate operation. Grate combustion requires understanding of heat transfer, gas flows, chemical reactions, and fuel properties. Ignition and combustion are therefore influenced by many different factors, the effect of which is very complex.

References

Caillat, S., Vakkilainen, E., 2013. Large-scale biomass combustion plants: an overview (Chapter 8). In: Rosendahl, L. (Ed.), Biomass Combustion Science, Technology and Engineering. Woodhead Publishing Series in Energy, London, pp. 274–296. ISBN 9780857091314.

Eriksson, L., Ingman, R., 2001. Recommendations for conversions of grate fired boilers to fluidising beds (Anvisningar för konvertering av rosterpannor till fluidiserad bäddteknik). Technical Report, SVF-725 Värmeforsk, Stockholm (Sweden), ISSN 0282-3772. 30 p (in Swedish).

Horttanainen, M., Saastamoinen, J., and Sarkomaa, P., 2000. Ignition and flame spread in fixed beds of wood particles. INFUB, Conference on Industrial Furnaces and Boilers, Porto, 12 p.

Huhtinen, M., 2000. Combustion of bark. In: Gullichsen, J., Fogelholm, C.J., (Series Eds.) Chapter 15 in Book 6, Chemical Pulping. Finnish Paper Engineers' Association and TAPPI, ISBN 9525216063.

Kortela, U., Marttinen, A., 1985. Modelling, identification and control of a grate boiler. Proceedings of the American Control Conference, pp. 544–549.

Ostendorf, F.J., 1986. Ignifluid—feuerung: kombination aus rost und wirklichkeit. (Ignifluid-firing: combined grate and fluidized bed). Brenstoff Wärme Kraft. 38 (5), 201–205 (in German).

Päällysaho, J., 2009, Arinakattilan lämmönsiirron laskentamallin kehittäminen (Development of heat transfer model for grate boiler furnace). M. Sc. thesis Lappeenranta University of Technology, 86 p (in Finnish).

Poltto- ja palaminen (Firing and combustion). In: Raiko, R., Kurki-Suonio, I., Saastamoinen, J., Hupa, M. (Eds.), Jyväskylä: Teknillisten Tieteiden Akatemia (TTA). Gummerus Kirjapaino Oy, 629 p. ISBN 9516664482 (in Finnish).

Thomas, H.-J., 1975, Thermishe Kraftanlagen (Thermal Powerplants). In German, Berlin, Springer, 386 p. ISBN 3540067795.

Yin, C., Rosendahl, L.A., Kær, S.K., 2008. Grate-firing of biomass for heat and power production. Prog. Energy Combust Sci. 34, 725–754.

10

FLUIDIZED BED BOILERS FOR BIOMASS

Since the late 1980s, fluidized bed boilers have emerged to dominate grate-fired boilers for sizes over $10\,MW_{th}$ (Hupa, 2005) (Eriksson and Ingman, 2001). Large circulating fluidized bed (CFB) boilers are challenging pulverized coal-fired (PCF) boilers in the largest utility boilers (Giglio, 2012). In particular, fuel flexibility, the possibility to cofire biomass, and the cost of carbon dioxide emissions are driving utilities toward adopting CFB instead of PCF boilers (Van Dijen et al., 2005)

Fluidized bed reactors have been used in petrochemistry and coal gasification since the 1930s. The first commercial applications for the combustion of solid fuels took place in the 1970s.

At first, fluidized bed firing gained foothold in Scandinavia, firing biomass, and in North America, firing bituminous coal. Since then, fluidized bed combustion has become widely accepted for the combustion of various solid fuels. Fluidized beds are very suitable for the combustion of low-grade fuels with high moisture or ash content (Atimtay and Topal, 2004). These are normally difficult to burn using other combustion methods. The benefits of fluidized bed combustion are the possibility to use several different fuels simultaneously, simple and cheap sulfur removal by injecting limestone into the furnace, high combustion efficiency, and low NOx emissions.

The two main types of fluidized bed combustion are: the bubbling fluidized bed boiler (BFB) and the CFB boiler. The BFB has a lower own energy consumption and a lower unit price up to about 100–300 MW_{th}. The CFB has a better environmental performance. Typically the BFB is used in applications of less than 100 MWth and the CFB from 50 MWel. The CFB is also better suited for burning coal (Van Dijen et al., 2005).

Each fluidized boiler has an individually optimized design for a certain fuel mixture and a unique main steam pressure and temperature. Therefore the order and placement of heat exchanger surfaces may vary in different boilers. For example, a part of the evaporator is in the convective section after the solids separator in a boiler designed for low steam pressure and moist fuel such as bark. For high steam pressure boilers, some superheater surfaces are usually located in the furnace.

Solid biofuels are not easy fuels to burn. They have a high moisture content and thus a low heating value. Even though the ash content is often low, solid biofuels frequently contain highly undesirable ash compounds (chlorine and potassium) that cause corrosion: In addition, the resulting ash often has a low melting temperature. However, fluidized bed boilers are ideal for handling these difficulties as they are not too sensitive to changes in the fuel composition and operate at a low steady temperature. The bed heat capacity helps to dampen the changes in fuel quality.

Typically biofuels are easy to volatilize. They have low sulfur but a medium to high nitrogen content. Handling biomass sulfur emissions has proven easy but fluidized bed boiler technology is still seeking methods to achieve lower NOx levels. Fluidized bed boilers have almost replaced all other types of boilers for the combustion of solid biofuels when the boiler size is in excess of 50 MWth.

10.1 Theory of Fluidized Bed Combustion

In fluidized bed combustion, air is blown through a layer of fine sand and ash called the bed or fluidized bed. As air velocity increases, the bed becomes fully suspended and starts to behave like fluid. Increasing the furnace temperature further increases the gas velocities. The biofuel is then mixed with hot fine solids at the bottom of the furnace. Burning of the biofuel is facilitated by continuous contact with hot solid particles. The mass fraction of biofuel particles is typically about 1–5% by weight of all bed solids. Because of the thermal capacity of the solids forming the bed, drying is fast and combustion occurs at a relatively fixed temperature.

Fig. 10.1 shows the operating principle of a fluidized bed. It illustrates the behavior of a bed of sand particles as a gas flows through the bed at different velocities, and the gas pressure drop through the bed as a function of gas velocity. For a fixed bed, the gas pressure drop is proportional to the square of the velocity. As velocity increases, the bed becomes fluidized. The velocity at which this transition occurs is the minimum fluidizing velocity, U_{mf}. At this point the drag force to particles in the fluidized bed equals the sum of forces acting on the particles. The minimum fluidizing velocity depends on many factors, including particle diameter, gas and particle density, particle

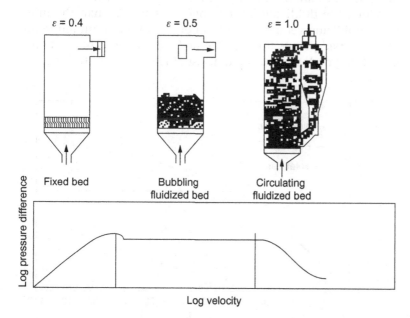

Figure 10.1 Different types of fluidization: left, fixed bed; center, bubbling fluidized bed; right, circulating fluidized bed.

shape, gas viscosity, and bed void fraction. The following formula calculates the minimum fluidizing velocity:

$$U_{mf} = \frac{\mu_s}{d_p \rho_s}\left[\sqrt{33.7^2 + 0.0408\frac{d_p \rho_p (\rho_g - \rho_s)g}{\mu_g^2}} - 33.7\right] \qquad (10.1)$$

where

μ_g is dynamic viscosity
d_p is particle diameter
ρ_g is gas density
ρ_p is particle density
g is acceleration of gravity

At velocities above U_{mf}, the pressure drop through the bed remains constant and equals the weight of solids per unit area as the drag forces on the particles barely overcome gravitational forces. The following equation shows the pressure drop

$$\Delta p = (\rho_g - \rho_s)(1 - \varepsilon)gH \qquad (10.2)$$

where

ρ_g is gas density
ρ_p is particle density
ε is the ratio of empty volume in the bed
g is acceleration of gravity.

Fig. 10.2 shows the gas pressure drop through a fluidized bed and the state of fluidization versus gas velocity.

When the fluidizing velocity becomes higher than the minimum fluidizing velocity, the gas in excess of that required to fluidize the bed passes through the bed as bubbles. This system is a bubbling bed, and boilers that use this system are BFB boilers.

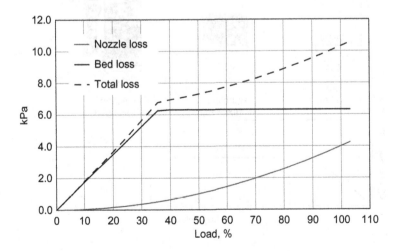

Figure 10.2 Gas pressure drop through an example fluidized bed and state of fluidization versus gas velocity expressed as % load.

The BFB has a modest solids mixing rate and a low solids entrainment rate into the flue gas. The bubbling bed has a clear, visible surface level where the bed ends and the freeboard begins. As the fluidizing velocity increases, the fluidized bed surface becomes more diffuse and solids entrainment increases. At about 2 m/s with a typical BFB the clearly defined bed surface no longer exists, and a recycling of the entrained material to the bed is necessary to maintain the bed inventory. After doubling the lower furnace's free velocity, and often changing to smaller diameter particles, the fluidized bed with these characteristics is called a CFB. The name is derived from the high amount of entrained solids. The operation then requires a particle separator and circulation back to the furnace of the separated bed material to keep the bed inventory. Boilers that use this system are CFB boilers.

Gravity, buoyant force, and a drag force by the fluidizing medium influence a solid particle in a fluidized bed, Fig. 10.3. If air is the fluidizing medium, the effect of the buoyant force is negligible. Forces that influence the particles can be calculated using the following formulas:

Drag force:

$$F_d = C_d \frac{\pi (d_p)^2}{4} \frac{1}{2} \rho_g U^2 \qquad (10.3)$$

where
  C_d is the drag coefficient
  d_p is the particle diameter
  U is the fluidizing velocity
  Buoyant force:

$$F_b = \frac{\pi (d_p)^3}{6} \rho_g g \qquad (10.4)$$

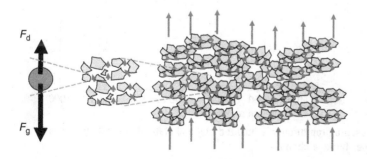

Figure 10.3 Bubbling fluidized bed: left, action of forces on a particle; right, clustering of particles and appearance of voids (bubbles).

Gravity force:

$$F_g = \frac{\pi (d_p)^3}{6} \rho_p g \qquad (10.5)$$

In a fluidized system the force balance is:

$$F_d + F_b = F_g \qquad (10.6)$$

The terminal velocity, U_t, is:

$$U_t = \sqrt{\frac{4}{3} \frac{d_p (\rho_p - \rho_g)}{\rho_g C_d} g} \qquad (10.7)$$

As the fluidizing velocity becomes higher than the terminal velocity, U_t, the solids start to separate from the bed. Fig. 10.4 illustrates the phenomenon. Fluidization can be affected not only by particle size but also by their distribution, density, shape, fluidization air speed, and bed density distribution.

An advantage of the fluidized bed technique is the good heat transfer between the gas and heat transfer surface. The heat transfer between the bed and the heat transfer surface rises rapidly when the velocity of gas exceeds the minimum fluidizing velocity. The explanation for this is that the hot bed material comes continuously into contact with the heat exchanger. Each hot bed material particle coming into contact releases heat to

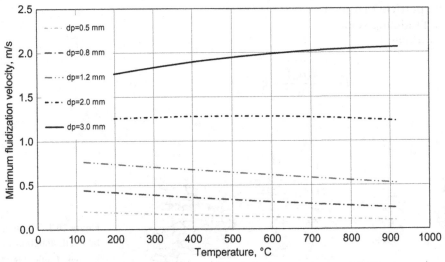

Figure 10.4 Terminal speed at gas pressure drop through a fluidized bed and state of fluidization versus gas velocity when temperature of lower furnace changes.

the boiler tube surface. As a steady stream of new hot particles comes into contact, due to the mixing of the fluid bed, high heat transfer rates are possible.

10.2 Bubbling Fluidized Bed Combustion

Fig. 10.5 shows a BFBboiler, and Table 10.1 presents typical operating parameters for a BFB boiler. The BFB consists of a furnace, with superheaters hanging on top of the furnace and at the backpass, followed by the hottest economizer tubes. The backpass is the portion of the furnace enclosure where flue gases start flowing downward. In the BFB shown in Fig. 10.5 the first backpass is continued as the second backpass due to layout reasons. In the second backpass one sees the economizer and the air preheaters before the cooled flue gases are drawn to the electrostatic precipitator (not shown). The height of the fluidized bed is 0.4−0.8 m resulting in a 6−12 kPa pressure drop over the bed. The bed material is sand and fuel

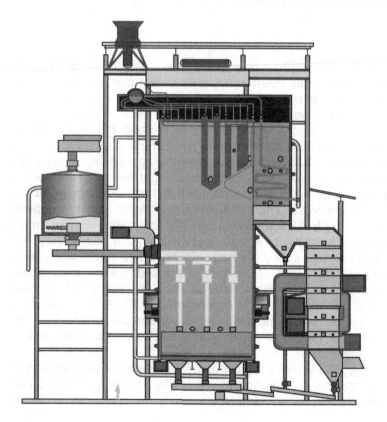

Figure 10.5 A bubbling fluidized bed (BFB) boiler (293 MW$_{th}$, 107 kg/s, 11.5 MPa, 541°C). Courtesy of Valmet.

Table 10.1 Typical operating parameters for a bubbling fluidized bed (BFB)

Volume heat load	MW/m^3	0.1—0.5
Cross-section heat load	MW/m^2	0.7—3
Pressure drop over the bed	kPa	6.0—12
Fluidizing velocity	m/s	1—3
Height of the bed	m	0.4—0.8
Bed material particle size	mm	0.5—1.5
Temperature of primary air	°C	20—400
Temperature of secondary air	°C	20—400
Bed temperature	°C	700—1000
Freeboard temperature	°C	700—1300
Excess air ratio	—	1.1—1.4
Density of bed	kg/m^3	1000—1500
Minimum load	%	30—40

Source: Adapted from Huhtinen, M., Kettunen, A., Nurminen, P., Pakkanen, H., 1999. höyrykattilatekniikka (Steam boiler technology). EDITA, Helsinki, 316 p. ISBN 951371327X (in Finnish) (Huhtinen et al., 1999).

ash. If sulfur capture is necessary, limestone is also fed into the furnace, and the reaction products (limestone) will form part of the bed material. The optimum bed material particle size is 0.5—2 mm when the fluidizing velocity is 0.7—2 m/s.

Crushed fuel is fed to the top of the bed. Fluidizing primary air is supplied to the bottom of the furnace through ducts or an air plenum followed by the distributing system. Heavier fuel particles dry and gasify in the bed. Light fuel particles such as peat and biomass tend to volatilize fast and burn above the bed in the freeboard. The bed temperature is 700—1000°C, depending on the fuel quality and load. The lower temperature limit is set by the ability to sustain sufficiently good combustion. The higher limit is set by the bed agglomeration, which stops fluidization. The high heat capacity of the bed material makes combusting low-grade fuels easy as they dry and volatilize easily when fuel particles make close contact with hot bed material. If swings or disturbances of the fuel properties or the feed rate occur, the high heat capacity of the bed material helps to even out the effects.

Fig. 10.5 shows a typical fuel feeding system for a BFB. From storage, fuel is conveyed to a day silo. This is to ensure a continuous fuel supply close to the boiler independent of the fuel loading and unloading operations. A screw discharges fuel from the bottom of the day bin into the fuel conveyor and feed chutes to the furnace. Air or screw feeders assist the fuel transport in the feed chutes, depending on the type of fuel. A rotary feeder is usually placed between the feed chute and the fuel conveyor to seal the fuel transport system from the hot flue gases in the furnace. Several feed chutes are necessary to distribute the fuel evenly across the full bed area. The fuel system typically causes most operational disturbances in bark boilers. A redundancy in the fuel feed system is often necessary to ensure good boiler availability.

During the boiler startup, the fluidized bed is first heated to 400–600°C with oil- or gas-fired startup burners to ensure the safe ignition of the solid biomass fuel (Kuo, 2005). The startup burners are typically above the bed, inside the bed, or in the primary air plenum or primary air duct. The fastest startup time and lowest oil or gas consumption result from heating the fluidizing or primary air in the air plenum or duct. Bed burners above are often necessary to provide load-carrying capability for situations where the solid fuel is unavailable for some reason.

The furnace enclosure is constructed of welded watercooled tubing (membrane wall) (Gavrilov, 2010). The tubes in the lower furnace are covered with refractory lining up to approximately 2 m from the grate level. The refractory is necessary to protect the tubes from erosion and assist in maintaining a sufficient bed temperature when fuels with a high moisture content are burned. The fluidizing primary air is distributed evenly over the whole gross section of the bed through nozzles in the grate. The grate is made of plate with tubes to cool it.

The turndown of a BFB is typically 30–40% of the maximum continuous rating (MCR) load with solid fuels. A load change capability of more than 5% can be achieved, depending on the dimensioning of the pressure part components. Lower loads require the use of support firing by oil or gas. The decrease in bed temperature primarily limits the minimum load. At continuous running of the minimum load the bed temperature should be above 700°C to ensure full combustion of all fuel. Fuel moisture therefore strongly influences the minimum load achievable without support fuel. Above MCR conditions, the maximum bed temperature, entrainment of bed material with flue gas, and increased emissions as well as unburned matter, limit the load. The BFB is very suitable for the combustion of fuels with a high volatile content, such as bark and wood wastes.

The BFB furnace consists of several distinct zones. The dense bed starts from primary/fluidization air introduction and

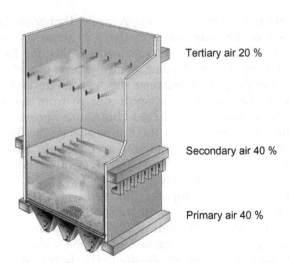

Tertiary air 20 %

Secondary air 40 %

Primary air 40 %

Figure 10.6 Bubbling fluidized bed (BFB) air distribution. Courtesy of Valmet Power.

extends to below the secondary air ports. Fuel fed to the furnace partly dries and volatilizes in this zone. There is good vertical mixing, but due to the dense bed height there are restrictions on fuel mixing in the horizontal direction. This requires careful consideration regarding the number of fuel feeding points. Most of the char combustion occurs in this dense bed. From top of the furnace close to the secondary ports is the freeboard. Volatilized gases burn, and combustion reactions are completed, in the freeboard (Saastamoinen, 2007). The freeboard has a very low concentration of sand and ash. Depending on the load rate, there is a zone between the dense bed and freeboard called the intermediate zone, which is not very thick. The bursting bubbles throw solids into this zone.

Air distribution and staging have a significant effect on combustion control and emissions. For low NOx emissions, tertiary air needs to be used. Tertiary air is injected fairly high up. Air introduced into the furnace can be labeled from bottom to top as primary, secondary, and tertiary air. In addition, cooling air is fed to cameras, burners, and other auxiliary equipment. Typically there is also some leakage air through, for example, sootblowing openings (Fig. 10.6).

10.3 Circulating Fluidized Bed Combustion

In the CFB combustion process the fluidization velocity is increased past the bubbling regime to the transport regime (Basu, 2006). High velocity causes a large fraction of the sand and

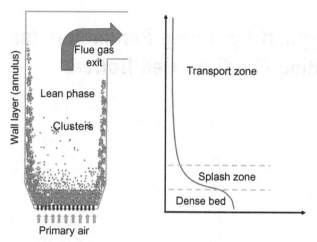

Figure 10.7 Circulating fluidized bed operating regimes and density gradient. After Gómez-Barea, A., Leckner, B., 2010. Modeling of biomass gasification in fluidized bed. Prog. Energy Combust. Sci. 36: 4, pp. 444—509 (Gómez-Barea and Leckner, 2010).

ash to become entrained with the flue gas stream (Monazam and Shadle, 2004). The furnace acts as the riser of a fast fluidized bed (Basu and Fraser, 1991). One needs some type of solid separator after the furnace to collect entrained particles. Collected particles are reinjected into the lower furnace. Some of the entrained particles hit the furnace walls, stopping and falling down as large particle clusters (Hua et al., 2004). This causes a gradual particle density profile between the lower furnace and the top furnace, Fig. 10.7. In practice, one needs a somewhat finer particle size to operate a CFB boiler than one needs for a BFB boiler. Because of the higher velocity, the cross-section area of a CFB furnace is much smaller than that of a BFB furnace of similar size. Because of their low uniform temperature, CFB boilers have a low emission of NOx. In addition, limestone reacts well and not much overdosing is needed for sulfur capture (Adánez et al., 1996).

In the CFB furnace the entrained solids form clusters of particles that can travel up or down the furnace. The formation of clusters causes good mixing and evens out temperatures in the furnace. Typically the temperatures of the gas and solids are practically uniform from the bottom of the furnace to the exit. Even though most CFB boilers burn coal, they are well suited to biomass burning and can easily handle biomasses such as bark, wood, and sludge (Frandsen et al., 1998). The best feature of the CFB furnace is that one can easily feed multiple fuels at the same time. Table 10.2 shows the typical operating parameters for CFB boilers.

Table 10.2 Typical Operating Parameters for Circulating Fluidized Bed Boilers

Volume heat load	MW/m^3	0.1 – 0.3
Cross-section heat load	MW/m^2	0.7–5
Total pressure drop	kPa	10–15
Bed material particle size	mm	0.1–0.5
Fly ash particle size	μm	<100
Bottom ash particle size	mm	0.6–10
Fluidizing velocity	m/s	3–10
Temperature of primary air	°C	20–400
Temperature of secondary air	°C	20–400
Bed temperature	°C	850–950
Temperature after the cyclone	°C	850–950
Excess air ratio	—	1.1–1.3
Density of bed	kg/m^3	10–100
Recirculation ratio	—	10–100
Minimum load	%	25–30

Source: Adapted from Huhtinen, M., Kettunen, A., Nurminen, P., Pakkanen, H., 1999. höyrykattilatekniikka (Steam boiler technology). EDITA, Helsinki, 316 p. ISBN 951371327X (in Finnish) (Huhtinen et al., 1999).

Fig. 10.8 shows a CFB boiler furnace. It differs from other biomass boiler types with regard to the solids separator, which is typically a cyclone (Kwauk and Li, 1999). The cyclone captures circulating bed material and unburned fuel particles and returns them to the lower part of the furnace through a loop seal (Lackermeier and Werther, 2002). After the cyclone, the flue gases flow through a normal arrangement of heat exchangers; superheaters, economizers, and air preheaters.

The CFB furnace walls are made of tubes welded together as membrane walls. Water evaporates inside the tubes so the tube surface temperature is 300–350°C. The lower furnace is typically tapered. The grate area is much smaller than the furnace cross area. Similarly to the lower furnace of a BFB boiler, a refractory lining is used up to 2–3 m in height from the grate (Goidich et al., 1999). When dimensioning the CFB furnace for given fuels and steam output, one must take into consideration:

- gas velocity (susceptibility to erosion increases with velocity)
- minimum burning time of fuel (temperature and air distribution to get an acceptable amount of unburned material)
- placement of the air ports (emission performance).

Figure 10.8 Schematic of a circulating fluidized bed (CFB) boiler. Courtesy of Valmet Power.

The fuel feeding in a CFB furnace can be done with a similar arrangement to the BFB furnace explained earlier. The amount of combustible material in the furnace is small: only 1−3% of the total mass of circulating material (Basu, 2015). In addition, fuel can be fed into the solids return in the loop seal. This can give good mixing and uniform distribution of fuel in a CFB boiler. Limestone for sulfur capture is typically fed pneumatically into the lower furnace through several feed points.

The two lower levels of the combustion air system in CFB boilers are similar to those of BFB boilers (Grammelis and Karakas, 2005). Primary air is blown through the nozzles located in the bottom of the furnace. The pressure of the primary air is 15−20 kPa, and the amount is 30−60% of the maximum total combustion air. Secondary air is supplied through several secondary air ports in one or two levels located 2−5 m from the bottom. The amount of secondary air can be regulated from 10% to 50% of the maximum total combustion air. A certain minimum pressure drop over the primary air nozzles is necessary to avoid uneven fluidization, improper temperature distribution in the bed, and back shifting of bed material into the wind box. At minimum load the primary air flow rate is about 50% of the MCR flow. Depending on the fuels used, the turndown of a CFB boiler is 25–40% of full load.

The bed operates at lower than stoichiometric air. When secondary air is injected, the air ratio approaches stoichiometric. In the CFB furnace the char particles are often transported upward and might make it all the way to the cyclones to be returned again to the lower furnace. Thus the char particles' combustion time can be fairly long, which means there is a low amount of unburned carbon even with fuels that are difficult to burn. Ash and sand are constantly fractured into smaller particles. Thus there is a stream of fine particles that escape from the cyclone with the fuel gas.

10.4 Fluidized Bed Operation

With solid biomass fuel, ash enters the furnace. Most of the biomass ash is removed through the separation of particles from the flue gases after they exit from the boiler. However, a significant amount of ash needs to be removed from the furnace bottom as bottom ash.

Fluidized bed operation requires that the fluidized bed particle size remains within the desired range. Depending on the fuel type, the fluidized bed average particle size increases or decreases.

10.4.1 Coarsening

Bed coarsening occurs with a rise in the proportion of large particles. That is, the average particle size in the bed increases. Coarsening is the result of bed particle agglomeration and the accumulation of char and unburned material in the fluidized bed. Coarsening disturbs the fluidization of the bed because larger particles start to accumulate in clusters, which reduces mixing. Coarsening causes just as much trouble in the BFB as in the CFB (Kaasalainen, 2009).

Coarsening of the bed changes the hydrodynamic behavior. It affects the bubble size of the bed, the bed mixing, and thus also the heat transfer in the furnace. In a CFB the bottom fraction of the circulating solids increases due to coarsening. As the particle size at the bottom increases, the density distribution in the furnace changes, often causing the heat transfer at the upper part of the furnace to decrease.

Coarsening decreases mixing and hinders fluidization. As fluidizing air is chaneled and fuel is distributed unevenly, the temperature distribution shows peaks, which lead to other usability problems such as the full or partial melting of ash. As a result of coarsening of the fluidized bed the heat transfer properties can be very uneven, which further increases the non-uniformity and causes occasional high temperatures. Poor mixing of the fluidized bed is detrimental to emission control because the air, fuel, and sorbents are mixed unevenly with each other, causing increases in all emissions.

The way we harvest solid biofuels can directly contribute to the amount of inert matter in the fuel. Particularly agricultural biofuels, stumps, and peat contain sand, stones, and other coarse noncombustible material, which accumulates from the fuel feed to the lower bed.

Alkalis such as potassium and sodium in the fuel will accumulate in the bed material. They decrease the melting point of the bed material and at high enough concentration will fuse or sinter the entire bed. As the sintering of the fluidized bed due to alkali accumulation occurs fairly fast after the required alkali level has been reached, fusing often surprises operators. Monitoring the alkali levels and continually changing the bed material prevents bed fusing. In addition, maintaining a good control of the bed temperature is essential. Typically the bed temperature should be below 900°C to avoid sintering. In practice it is enough to control the ratio of fluidizing air to total air flow to have sufficient fluidized bed temperature control.

For economical reasons it is impossible to completely remove impurities. No matter what the fuel vendors claim, a variety of unwanted large particles and impurities will inevitably end up in the furnace.

10.4.2 Removal of Bottom Ash

To slow down bed coarsening and maintain proper fluidization, it is important to renew the bed material at regular intervals. In the renewal of the bed material the coarse material and accumulated ash is removed from the bottom of the bed. The shape of the bottom of the furnace, location of discharge openings, and positioning of fluidization air jets affect the removal of coarse material. Removal of bed material helps to maintain the particle size distribution in the bed and a proper bed inventory (Basu, 2015).

Biomass ash from the fluidized bed has not typically sintered and is not spherical. Therefore it is often not very dense and has a high surface area. As coarse material tends to be found close to the bottom, its removal is done through ash chutes starting from openings in the bottom grate of the furnace. The bottom ash is transported with watercooled screws that help to cool it to an acceptable temperature. Bottom ash is then conveyed to an ash container. The bottom ash removal rate is controlled to maintain a constant bed material inventory in the furnace (Teir, 2004).

It is common for some metal objects, tools, wire, and other discarded objects to occasionally enter with the solid biofuel streams. As metals do not burn, they often form hard-to-remove clumps. Scrap metal, which accumulates in the bed, hinders fluidization and eventually causes problems not only to fluidization but bed temperature control and emissions. Often the only way to remove scrap metal is to completely shut down the boiler, remove the sand and manually remove these clumps.

Continually maintaining a correct level of bottom ash removal is important for solid biomass fuels that contain alkalis such as potassium and sodium. These can accumulate in the bed material and cause large-scale sintering.

10.4.3 Control of Particle Size in Bed

Depending on the evolution of the particle size distribution, new bed material with a suitable particle size is added to the bed. Typically the particle size is smaller than average. If too much additional fine material is added, then the upper furnace has a higher density of solids. The benefit is that a higher heat transfer coefficient is gained in most parts of the furnace. The drawback is that more fly ash is generated.

There is also the possibility of screening the bottom ash and returning the usable sand fraction back to the furnace. This is a good way of reducing the amount of new sand one has to purchase. The accumulation of agglomerating chemicals such as alkali should be taken into consideration when determining the maximum possible return rate. Some biofuels, such as green chips containing leaves and small branches, have a high alkali content. Some industrial residues (e.g., olive, plywood, or chipboard waste) contain high levels of alkali. Screening tends to return particles coated with alkali, which causes agglomerates with quartz. A solution is to use diabase or another low-quartz sand as bed material. With solid biofuels the new bed sand addition rate ranges from 1 to 4 kg/MWhpa. The amount of sand added is about 0.3–1% of the fuel mass flow.

10.4.4 Fuel Feeding

A proper fuel feeding system is an essential part of a well-functioning fluidized bed boiler. The fuel should be fed so that it is distributed evenly to the whole bottom region of the boiler. In BFB boilers the fuel is dropped from chutes in several feeding points to the bubbling bed and the mixing in the bed causes fuel to disperse evenly. In CFB boilers the fuel is fed by screws or other mechanical means though several feed point openings on the furnace walls to the refractory-lined lower furnace or mixed with return solids.

Depending on the boiler size, the fuel can be fed either from one wall or from two opposite walls. A higher number of feeding points gives more uniform distribution of fuel but is more expensive. One feed point in a CFB can serve $9-27\,m^2$ of bed area (Teir, 2004). This is significantly higher than for BFB boilers, where values less than $10\,m^2$ of bed area need to be used. The solid biomass fuel feed points for CFB boilers are typically located at the refractory-lined lower furnace. Some CFB boilers feed fuel to the return loop. In BFB boilers the solid biomass fuel is often dropped into the bubbling bed from high above.

A common arrangement is that fuel needs to be removed from flat-bottomed day silos with a rotating screw. Solid biomass fuels tend to be hard to remove from silos as they tend to bridge. Fuel is then dropped to a screw and conveyor arrangement that feeds fuel along the furnace walls. Dosing or stealing screws feed the fuel to rotary feeders. Their rotation controls the fuel flow to feeding chutes in each feeding point. Uneven feeding of solid biomass can be seen immediately, for example, in the furnace pressure deviation.

Solid biofuel particles behave differently, depending on their size. Small biomass particles tend to be carried with the gas and burn above the bed. Bigger biomass particles behave like coarse particles and sink. Volatilization of solid biomasses occurs close to the feeding point. Residual char burns in the bed.

Before one can start the solid biomass feed into a fluidized bed boiler, the bed temperature must be raised to $500-600°C$ so that the biomass ignites properly. After biomass feeding starts, the bed temperature can be rapidly increased. Initial bed heating is done using light oil or natural gas start-up burners. Usually the operating bed temperature is between $750°C$ and $900°C$. Control of the bed temperature is easy because of the high heat capacity of the bed.

If the properties of the fuel fed to a CFB boiler change drastically, then adjustments to the operation need to be done.

Air ratio and the bed inventory are the easiest things to change, but sometimes operability is improved by also altering the bed particle size or the size distribution.

10.4.5 Limestone Feeding

Limestone feeding can be used to enable SO_2 capture in the fluidized bed. In the fluidized bed the limestone particles will calcine to CaO and get porous. CaO can react with sulfur to sulfate. Limestone is typically transported by a bucket hopper from a loading hopper to a limestone silo. Limestone silos often have a conical bottom and a narrow but tall shape. Typically limestone can be injected into the fluidized bed by using gravity alone if the silo is properly positioned. Flow is often regulated by a rotating valve. The limestone particle size should not be too small as the fed particles need to stay in the furnace.

Sulfation of CaO is often hindered by the sulfate layer formed on the outer surface of the CaO particle. In practice one often finds an unsulfated core when ash is removed. CFB boilers require a Ca/S ratio of about 2 or less to achieve over 90% sulfur removal, whereas in BFB boilers the ratio is close to 3. Typically limestone is injected from one or two points into the lower part of the furnace.

10.5 Separation of Particles From Gas

In a CFB the particle size is small enough that some of the sand and ash is constantly transported from the furnace and needs to be returned to maintain the bed material inventory (Davidson, 2000). It should be noted that the majority of the particles separate into clusters on the walls and only a portion of the material in the furnace escapes. These particles are captured in a gas−solid separator and are recycled back to the furnace. Typically the capture is done by a cyclone-type device. The separation can be designed for a very high solid collection efficiency (>99%), so the CFB boiler backpass operates in a fairly clean state.

The dust emissions from a BFB or the fine dust from a CFB are collected by electrostatic precipitators or bag-house filters. Auxiliary equipment such as precipitators and filters is covered in more detail in Chapter 6, Auxiliary Equipment. Typically these are installed between the boiler exit and the stack. Collected fly ash is often pneumatically transported to the ash silo and from there to disposal.

10.5.1 Loop Seal

In CFB boilers the separated solids are returned to the furnace via the loop seal, Fig. 10.9. The weight of the returned solids provides a pressure seal against the pressure loss caused by the cyclone. Solids can be kept moving by gravity by injecting fluidizing air to the bottom of the loop seal. The loop seal prevents the flue gas in the furnace from short-circuiting to the cyclone. The loop seal has no moving parts and is simple and reliable to operate. Several alternative designs for a loop seal exist. The most typical is the N-type shown in Fig. 10.9. Another commonly used type is the L-type. All types use air to fluidize the returning solids column to provide pressure to compensate for the furnace pressure.

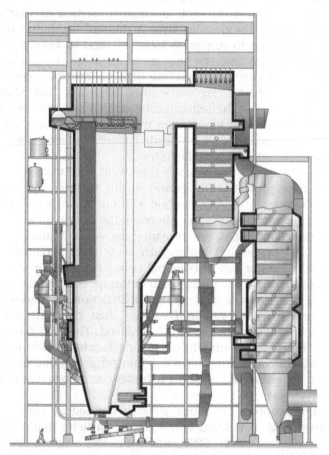

Figure 10.9 Return of bed material from the cyclone via a loop seal, fuel feeding, and bottom ash classification in a circulating fluidized bed (CFB) boiler.

The loop seal has been used to provide a feeding point for fuel and sorbent. Returning solids assist in fuel mixing and distribution in the furnace. With solid biomass fuels that volatilize fast, feeding fuel to the loop seal has not been popular as furnace side wall feeding provides better uniformity.

10.5.2 Cyclones

A CFB boiler typically has one or more high temperature cyclone-type solid separators. These cyclones can be built separately, Fig. 10.8, or integrated into the furnace. Small-diameter cyclones would have a higher collection efficiency, but costs favor building fewer and bigger cyclones.

The separate uncooled cyclones are typically constructed of steel plate. The inside is lined with abrasion-resistant refractory to provide erosion protection against separated particles. Between the steel and inner layer a thermal insulation refractory layer is used to minimize heat loss and to keep the steel plate temperature low. The amount of refractory in this type of cyclone is very large and therefore high maintenance costs and availability problems are envisioned. Starting and shutting down takes time as the temperature gradients in cyclones cannot be too high. Heat losses from the separate cyclones are significant. Expansion joints are needed. Over the years alternative cyclone constructions have been tried.

Cyclones made of membrane tube walls with water or steam cooling have proved costly. However, due to their low maintenance requirements, construction without an expansion joint and only a thin refractory have become the dominant solution.

Integrated cyclones are provided by some manufacturers. Fig. 10.9 shows a CFB boiler with a compact separator. The compact separator has a polygonal cross-section. The separator walls are made of flat pieces of membrane wall similar to the walls in the furnace. The panel construction decreases construction costs, but similar backpass dust concentrations to those in circular cyclones can be achieved. The CFB boiler can also be built with a multiinlet cyclone located inside the furnace. Integrated cyclones have a lighter refractory and are faster to start up compared with CFB boilers equipped with traditional cyclones.

Cyclone capture efficiency is often defined as cut-off size. The cut-off size is defined as the size of particles that are likely to be collected with 50% efficiency by a given cyclone. Increasing the inlet velocity to the cyclone improves capture efficiency as well as the pressure drop. A large cyclone diameter provides a smaller velocity gradient but the cost of the cyclone increases. The aim of cyclone design is to find the optimum parameters for efficiency and low cost.

10.5.3 U-Beam Particle Separators

An alternative option to cyclones is to use a primary particle separator that functions by impact force. If one places several rows of solid collectors consisting of U-beams so that the U faces the gas flow, then particles tend to hit the inside of the U and get separated. Typically in industrial applications there are 6–8 rows of U-tubes placed in a staggered arrangement. The advantage of U-beams is their small volume and the fact that they can be made from stainless steel. In a typical application one could find U-beams that are 152 mm wide by 178 mm deep. They are suspended by bolts though the furnace roof. To protect the beams from erosion a chromium oxide coating layer is used.

However, the capture efficiency of industrial U-beam separators is not always high and often additional dust separation needs to be done, e.g., by using multicyclones (Basu, 2015).

10.6 Heat Transfer in Fluidized Boilers

Solid biomass burning in a furnace generates hot flue gases. The furnace transfers heat from the hot combustion gases to water flowing inside the furnace wall tubes. This heat converts water to steam. In the gas–solid suspension of the fluidized bed furnace the solids serve firstly as additional heat capacity by alternately heating and cooling, and in the CFB boiler by exchanging heat when in contact with the furnace wall tubes (Al-Busoul, 2004). In typical fluidized bed furnaces the heat transfer between gas and solids is effective, due to good mixing.

In many boiler types the heat transfer in the furnace is dominated by the radiative heat transfer. This is also the case for BFB boilers. In CFB boilers the flue gas to furnace wall heat transfer essentially depends on the behavior of the suspension. Particle gas suspension flow behavior is dynamic and depends on the particle size distribution, which varies in different parts of the combustion chamber and depends on the size of the boiler. Therefore, a universal and accurate method for estimating the heat transfer to the CFB furnace walls has not yet been developed, even though good engineering approximations exist (Table 10.3).

Table 10.3 Typical heat transfer coefficients in fluidized beds

Heat Transfer Surface	Overall Heat Transfer Coefficient (W/m^2K)	
	BFB	CFB
Bottom furnace	80—250	80—250
Furnace walls	80—250	80—250
Furnace internals	50—250	50—250
Cyclone		40—60
Return leg	300—500	300—500
Superheaters, reheaters	200—300	200—300
Economizers	200—300	200—300
Air heaters	15—30	15—30

BFB, bubbling fluidized bed; *CFB*, circulating fluidized bed.

What makes fluidized bed heat transfer challenging is that solids and gas volume fractions, velocities, and temperatures have large gradients from the vicinity of the heat transfer surfaces both temporally and locally. Therefore the developed heat transfer correlations are based on averaged values over time and across the wall surface of the furnace. For example, suspension density is generally determined by means of the vertical pressure gradients in the furnace. In the vicinity of heat transfer surfaces the concentration of solids may actually be something completely different than the average concentration specified across the surface. The correlations are often based on furnace temperature measurements made at a specified distance from the wall. This means that the published correlations often apply to a specific temperature, suspension density profile, and furnace size (Viljanen, 2009).

10.6.1 Fluidized Bed Furnace Heat Transfer

We often discuss the lower furnace and upper furnace heat transfer separately. As the lower furnace has refractory, the furnace heat transfer correlations often apply only to the upper furnace. The fluidized bed is always dynamic and the suspension density varies as regions with hardly any solids alternate with regions with a higher suspension. Examples are bubbles in the BFB, where the density is much lower than around them, solid clusters in the CFB upper furnace, and backflow along the CFB furnace walls.

The vicinity of the furnace walls tends to become saturated with clusters of particles that move down along the wall surface for a brief period before they are torn apart. As these clusters travel along the wall, they cool (both gas and solids) and heat the wall tubes. There is often a clear high density boundary layer in the vicinity of the walls. At the same time a temperature gradient is formed. The temperature of the gas and the solid material increases from the wall toward the center of the furnace (Luan et al., 1999). The higher the amount of solids that are separated onto the walls, the higher is the heat flux.

For BFBs one can calculate the furnace heat transfer based on typical radiative furnace calculations (see Chapter 5: Thermal Design of Boiler Parts). Basu (2015) gives a rough first estimate for the average furnace heat transfer of typical commercial CFB furnaces as

$$h_f = 68 + 12u_f \tag{10.8}$$

where
  u_f is the superficial gas velocity, m/s

10.6.2 Heat Transfer at Part-Load Operation

Density at the lower furnace is always high. Down to about 60% MRC load the bed temperature remains fairly constant. For CFBs the suspension density at the upper furnace starts to be very thin at about 70% of MCR load. Then particle to wall convection becomes low and radiative transfer dominates (Xie et al., 2003). No matter how high the load is, the bed density at the lower section of the boiler is always very high. But as the load decreases, the density at the upper section is decreased and the bed is diluted. At 70% of full load, particle concentration at the upper section of the furnace, where most of the heat transfer surfaces are located, is weak. Therefore, convection heat transfer becomes weak, while the radiation process becomes dominant (Shi et al., 1998). If the load is reduced to 40%, the CFB furnace operates like a BFB boiler, and above the bed region radiation dominates as heat transfer.

For BFBs the bed temperature is also fairly constant down to about 60% MCR load. Heat transfer is dominated by radiation. Part load behavior is similar to other biomass boilers where the location of combustion (% above the fluid bed), radiation heat transfer, and flue gas mass flow allow partial load calculation using simple heat transfer methods.

10.6.3 Furnace Design

Furnace design of any fluidized bed boiler follows the principles of furnace design for all boilers. The first step is to determine the cross-sectional area of the furnace. For both BFB and CFB boilers this is typically done by selecting the furnace cross-sectional loading (the hearth heat release rate or HHRR). Based on the required furnace area, a height is then chosen. For more details, see Chapter 5, Thermal Design of Boiler Parts.

CFB boilers can contain additional furnace heat transfer surfaces. Typical ones are wing wall superheaters. The wing wall superheater is a panel-type superheater that extends from the front wall through the roof, Fig. 10.8. It has become popular, especially in CFB applications. The tube material is either carbon steel welded together with welded erosion protection or special omega-type tube. Instead of a superheater, a similar construction can be used to provide a more evaporative heat transfer surface, Fig. 10.9.

10.6.4 External Fluid Bed Heat Transfer Surfaces

BFB and CFB boilers burning biomass typically have no heat transfer surfaces at the dense part of the fluidized bed. Several BFB boilers that burn coal have been built using fluid bed heat transfer surfaces to limit the lower furnace bed temperature. For biomass these are not needed. Even for coal the use is limited as bed vibration and movement cause large stresses and erosion to the heat transfer tubes.

The newest CFB boilers place a small BFB with heat exchangers at the cyclone return flow. One can also utilize the return flow along the furnace wall.

10.7 Fluidized Bed Boiler Retrofits

A large number of old boilers are converted to fluidized bed boilers every year. Conversion is typically cheaper than buying a completely new boiler.

The main disadvantage of boiler conversion is that steam pressure and temperature, and often also the flue gas exit temperature, remain the same as in the original boiler. Therefore increasing electricity production or improving efficiency is not possible. Often the existing furnace is fairly small and space is limited, so many emissions might be slightly higher than in new boilers. One also has to take into account the fact that the remaining equipment is still old, so its technical lifetime is restricted and maintenance costs are typically higher than for completely new equipment.

In addition to a fluidized bed retrofit costing about half of the cost of a new boiler, the main advantage is the short project schedule. Often in about a year from the project startup one can enjoy the benefits. The actual erection time is faster as the boiler house remains essentially as it was. Also the permit process for retrofits is often simpler and more straightforward. Typically in addition to lower emissions, one of the advantages is a wider range of fuels, including moister fuels such as sludges (Latva-Somppi, 1998).

References

Adánez, J., de Diego, L.F., Gayán, P., Armesto, L., Cabanillas, A., 1996. Modelling of sulfur retention in circulating fluidized bed combustors. Fuel. 75 (3), 262–270.

Al-Busoul, M.A., 2004. Bed-to-surface heat transfer in a circulating fluidized bed. Heat Mass Trans. 38 (4-5), 295–299.

Atimtay, A.T., Topal, H., 2004. Co-combustion of olive cake with lignite coal in a circulating fluidized bed. Fuel. 83 (7-8), 859–867.

Basu, P., 2006. Combustion and Gasification in Fluidized Beds. CRC Press, 345 p. ISBN 0849333962.

Basu, P., 2015. Circulating Fluidized Bed Boilers: Design, Operation and Maintenance. Springer, 366 p. ISBN 9783319061726.

Basu, P., Fraser, S.A., 1991. Circulating Fluidized Bed Boilers. CRC Press, 345 p. ISBN 075069226X.

Davidson, J.F., 2000. Circulating fluidised bed hydrodynamics. Powder Technol. 113 (3), 249–260.

Eriksson, L., Ingman, R., 2001. Recommendations for conversions of grate fired boilers to fluidising beds (Anvisningar för konvertering av rosterpannor till fluidiserad bäddteknik). Technical Report, SVF-725 Värmeforsk, Stockholm (Sweden), ISSN 0282-3772. 30 p (in Swedish).

Frandsen, F., van der Lans, R., Pedersen, L.S., Nielsen, H.P., Hansen, L., Lin, W., et al., 1998. Firing in Fluid Beds and Burners. Technical report, CHEC-R—9805, Danmarks Tekniske Univ., Lyngby (Denmark). Institute for Kemiteknik. 91 p.

Gavrilov, A., 2010. Modeling of water/steam circulation in circulating fluidized bed boiler. M.Sc. Theses, Lappeenranta University of Technology, 92 p.

Giglio, R., 2012. CFB set to challenge PC for utility-scale USC installations. Power Engineering International. 16–21, January 2012.

Goidich, S.J., Hyppanen, T., Kauppinen, K., 1999. CFB boiler design and operation using The INTREX™ heat exchanger. 6 International Conference on Circulating Fluidized Beds, August 22–27, 1999, Würzburg, Germany, 6 p.

Gómez-Barea, A., Leckner, B., 2010. Modeling of biomass gasification in fluidized bed. Prog. Energy Combust. Sci. 36 (4), 444–509.

Grammelis, P., Karakas, E., 2005. Biomass combustion modeling in fluidized beds. Energy Fuels. 19 (1), 292–297.

Hua, Y., Flamant, G., Lu, J., Gauthier, D., 2004. Modelling of axial and radial solid segregation in a CFB boiler. Chem. Eng. Process. 43 (8), 971–978.

Huhtinen, M., Kettunen, A., Nurminen, P., Pakkanen, H., 1999. Höyrykattilatekniikka (Steam boiler technology). EDITA, Helsinki, 316 p. ISBN 951371327X (in Finnish).

Hupa, M., 2005. Fluidized bed combustion. Politecnico di Milano Lecture series, Options for Energy Recovery from Municipal Solid Waste, January 31st–February 2nd, 2005, 41 p.

Hyppänen, T and Raiko, R., 2002. Leijupoltto (Fluidized bed combustion). In: Raiko, R., Kurki-Suonio, I., Saastamoinen, J., Hupa, M., (Eds.) Poltto ja palaminen. Jyväskylä, International Flame Research Foundation, Suomen kansallinen osasto, pp. 490–521. ISBN 9516666043 (In Finnish).

Kaasalainen, J., 2009. Diagnostiikkamenetelmät höyrykattilan leijukerroksen tilan ja käytettävyysongelmien tunnistamiseksi (Bed diagnostic methods for recognizing the state and usability problems in fluidized bed boilers). M.Sc. Thesis, Lappeenranta University of Technology, Energy Technology, Lappeenranta, 112 p (in Finnish).

Kuo, K.K.-Y., 2005. Principles of Combustion. Wiley-Interscience, 736 p. ISBN 0471046892.

Kwauk, M., Li, J.H. (Eds.), 1999. Circulating Fluidized Beds (CFB) Past, Present And Future. Pergamon Press, OxfordKwauk, M., Li, J.H. (Eds.), 1999. Circulating Fluidized Beds (CFB) Past, Present And Future. Pergamon Press, Oxford, 272 p. ISBN 9787030061119.

Lackermeier, U., Werther, J., 2002. Flow phenomena in the exit zone of a circulating fluidized bed. Chem. Eng. Process. 41 (9), 771–783.

Latva-Somppi, J., 1998. Experimental Studies on Pulp and Paper Mill Sludge Ash Behavior on Fluidized Bed Combustors. Ph.D. Thesis, VTT Publications 336, Technical research centre, VTT, Espoo, 155 p. ISBN 9513852148.

Luan, W., Lim, C.J., Brereton, C.M.H., Bowen, B.D., Grace, J.R., 1999. Experimental and theoretical study of total and radiative heat transfer in circulating fluidized beds. Chem. Eng. Sci. 54 (17), 3749–3764.

Monazam, E.R., Shadle, L.J., 2004. A transient method for characterizing flow regimes in a circulating fluid bed. Powder Technol. 139 (1), 89–97.

Saastamoinen, J., 2007. Simplified model for calculation of devolatilization in fluidized beds. Fuel. 85 (17–18), 2388–2395.

Shi, D., Nicolai, R., Reh, L., 1998. Wall-to-bed heat transfer in circulating fluidized beds. Chem. Eng. Process. 37 (4), 287–293.

Teir, S., 2004. Steam Boiler Technology. 2nd ed. Energy Engineering and Environmental Protection publications, Helsinki University of Technology, Department of Mechanical Engineering, 215 p. ISBN 9512267594.

Van Dijen, F., Savat, P., Vanormelingen, J., Sablon, H., 2005. Ultra supercritical pulverised fuel combustion versus supercritical circulating fluidized bed combustion: is ultra supercritical circulating fluidized bed combustion on top? VGB PowerTech. 85 (11), 64–66.

Viljanen, J., 2009, Kiertoleijukattilan lämmönsiirto tulipesässä ja leijukerroslämmönsiirtimessä (Heat transfer in furnace and fluidized bed heat exchanger of a circulating fluidized bed boiler). M.Sc. Thesis, Lappeenranta University of Technology, Energy Technology, Lappeenranta, 94 p. (in Finnish).

Xie, D., Bowen, B.D., Grace, J.R., Lim, C.J., 2003. Two-dimensional model of heat transfer in circulating fluidized beds. Part II: heat transfer in a high density CFB and sensitivity analysis. Int. J. Heat Mass Trans. 46 (12), 2193–2205.

11

RECOVERY BOILER

Kraft recovery boilers are the biggest and largest producers of steam from biomass. They burn the organic residue from chemical pulp manufacture. About half of the mass of wood input is converted to pulp for paper making and the rest is burned. Kraft recovery boilers burn about a quarter of all modern biomass in the world.

Steam Generation from Biomass. DOI: http://dx.doi.org/10.1016/B978-0-12-804389-9.00011-3

11.1 Principles of Kraft Recovery

Spent cooking chemicals and dissolved organics are separated from pulp during washing. At first this black, alkaline liquor was dumped. Chemical recovery systems were used earlier, but it was in the 1930s and 1940s that a modern type of regeneration of spent liquor was widely adopted. The invention of new types of equipment and an increase in mill size led to a favorable economic situation. It was cheaper to process black liquor than to buy new chemicals.

Recovery of black liquor has other advantages. Concentrated black liquor can, when burned, produce energy for the generation of steam and electricity. In the most modern pulp mills, this energy is more than sufficient to cover all internal use.

The main kraft recovery unit operations, Fig. 11.1, are divided into recovery and fiberline. In recovery the unit operations are: evaporation of black liquor to reduce the amount of heat for vaporization of water and combustion of black liquor in the recovery boiler furnace including the formation of sodium sulfide and sodium carbonate, the causticizing of sodium carbonate to sodium hydroxide, and the regeneration of lime mud in a lime kiln to produce calcium oxide.

There are other minor operations to ensure continuous operation of the recovery cycle. Soap in the black liquor can be removed and tall oil produced. Control of the sodium— sulfate balance is done by the addition of makeup chemicals such as

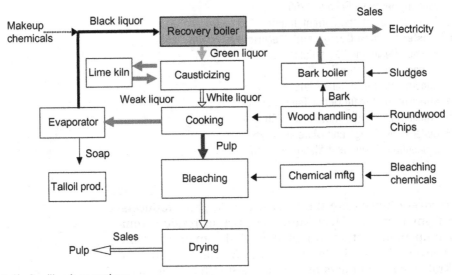

Figure 11.1 Kraft mill unit operations.

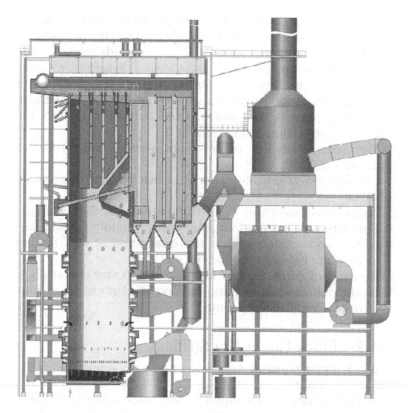

sodium sulfate to mix in the tank or by the removal of recovery boiler fly ash. Removal of the recovery boiler fly ash removes mostly sodium and sulfur, but also serves as a purge for chloride and potassium (Fujisaki et al. 2003). Buildup of nonprocess elements is prevented by the disposal of dregs and grits during causticizing. Malodorous gases are processed by combustion in the recovery boiler or lime kiln. In some modern and closed mills, chloride and potassium removal processes are employed. With additional closure, new internal chemical manufacturing methods are sometimes applied (Fig. 11.2).

11.1.1 Function of the Recovery Boiler

The recovery boiler burns concentrated black liquor. In addition to water this contains organic dissolved wood residue and chemicals used to remove lignin from cellulose fibers. When the organic portion is burned, heat is produced. The heat from the recovery boiler generates high-pressure steam. Steam expansion in a turbine generates electricity. Low-pressure steam exiting the turbine is used for process heating applications in the pulp mill.

Black liquor contains sulfur. Sulfur compounds released during combustion are captured when sodium reacts to produce sodium sulfate and carbonate. Optimum process conditions are needed to reduce sulfur gas emissions. The recovery boiler also serves as a process component. Used pulping chemicals are regenerated to sodium sulfide (Na_2S).

The recovery boiler process has several unit processes (Vakkilainen, 2005):

1. Combustion of organic material in black liquor to generate steam
2. Reduction of inorganic sulfur compounds to sodium sulfide
3. Production of molten inorganic flow of mainly sodium carbonate and sodium sulfide and dissolution of said flow to weak white liquor to produce green liquor
4. Recovery of inorganic dust from flue gas to save chemicals
5. Production of sodium fume to capture combustion residue of released sulfur compounds.

11.1.2 Black Liquor Dry Solids

Black liquor contains very fine organics, various chemicals, and water as a mixture. Black liquor dry solids are defined as a mass ratio of dried black liquor to wet black liquor before drying. If the water content in black liquor is above 80%, it has a negative net heating value, Fig. 11.3. In that case, all heat from the combustion of organics in black liquor would be spent evaporating the water it contains.

Black liquor dry solids have always been limited by the ability of available evaporation technology to handle highly viscous liquor (Holmlund and Parviainen, 2000). The design of dry solids for greenfield mills has been either 80% or 85% dry solids. In Asia and South America 80% (and before that 75%) dry solids have been in use. In Scandinavia and Europe 85% (and before that 80%) have been in use.

11.1.3 High-Temperature and Pressure-Recovery Boiler

In the beginning the development of the recovery boiler's main steam pressure and temperature was rapid, Fig. 11.4. By 1955, not even 20 years after recovery boilers were first created, the highest steam pressures were 10.0 MPa with a temperature

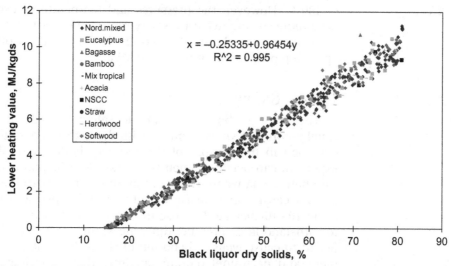

Figure 11.3 Net heating values of typical kraft liquors at various concentrations. From Vakkilainen, Esa K., 2000, Estimation of elemental composition from proximate analysis of black liquor. Paperi ja Puu - Paper and Timber, Vol. 82, No. 7, pp. 450–454.

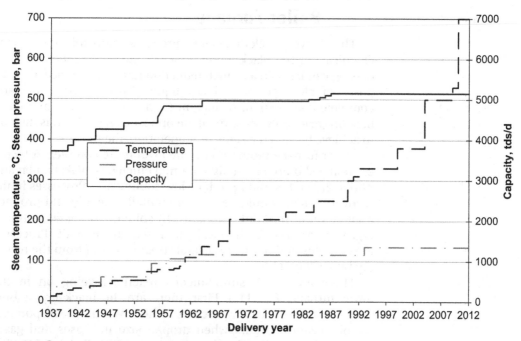

Figure 11.4 Development of recovery boiler pressure, temperature, and capacity.

of 480°C. However, the pressures and temperatures used then were reduced somewhat for safety reasons (McCarthy, 1968). By 1980 there were about 700 recovery boilers in the world with operating pressure at or above 8.0 MPa.

11.1.4 Safety

The recovery boiler has unique safety hazards. If even a small amount of water is mixed with the high-temperature solid char bed in the bottom of the recovery boiler, a smelt-water explosion can occur. A smelt-water explosion is purely a physical phenomena where water evaporates fast. This sudden evaporation causes an increase in volume and a pressure wave of some 10–100,000 Pa. Because the furnace walls cannot be built to withstand large pressure differences, this force is usually sufficient to cause walls to bend out of shape. The safety of equipment and personnel requires an immediate shutdown of the recovery boiler if there is any possibility that water has entered the furnace.

11.2 Chemical Processes in the Recovery Boiler Furnace

The recovery boiler efficiently processes inorganic and organic chemicals in the black liquor. Efficient inorganic chemicals processing can be seen as a high reduction rate. The furnace also disposes of all organics in black liquor. This means stable and complete combustion. Reduction (removal of oxygen) and combustion (reaction with oxygen) are opposite reactions. It is difficult to achieve both in the same unit operation, the furnace.

Other furnace requirements are even more complex. A recovery boiler should have a high thermal efficiency (Vakkilainen and Ahtila, 2011). It should produce low-fouling ash. Processes in the recovery boiler should be environmentally friendly and produce a low level of harmful emissions. In spite of successes optimizing, recovery boiler chemical processes are difficult. Processes are complex and there are several streams to and from the recovery boiler (Fig. 11.5).

There are many simultaneous reactions going on in the lower furnace, Fig. 11.5. First, there are the black liquor burning processes. Drying occurs when water is evaporated, Devolatilization occurs when droplet size increases and gases generated inside the droplet are released. Finally, char burning takes place when carbon is burned off. In the lower part of the

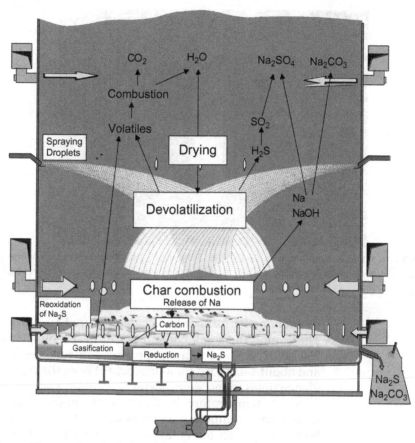

Figure 11.5 Some of the reactions in the lower furnace.

furnace there are char bed reactions. These consist mostly of inorganic salt, especially smelt reactions. In the upper furnace there is combustion of volatiles. Almost all other combustion reactions are concluded. Sodium sulfate and carbonate fume formation along with other aerosol reactions take place. There are a multitude of chemical reactions that take place in the recovery boiler. The best way to study them is to look at them main component by main component.

11.2.1 Smelt

Smelt is the product of inorganic reactions in the recovery furnace. At the same time as the carbon is consumed, the residual inorganic portion melts. Inorganics flow out of the furnace through smelt spouts, Fig.11.6 The amount of smelt inside the

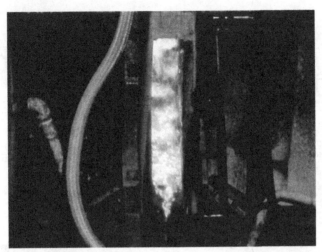

Figure 11.6 Smelt flow from the char bed. From Saviharju, K., Pynnönen, P., 2003. Soodakattilan keon hallinta (Control of Kraft recovery boiler char bed). Konemestaripäivä, 23.1.2003, Oulu, Finnish recovery boiler users association, 16 p (in Finnish) (Saviharju and Pynnönen, 2003).

recovery boiler furnace has been measured by Kelly et al. (1981). They found the smelt content per furnace unit area to be about 250 kg/m^2 for a decanting Combustion Engineering (CE) unit and about 140 kg/m^2 for a Babcock&Wilcox (B&W) unit. The residence times found were 44 and 25 minutes respectively.

Smelt temperature is about 100°C higher than the initial deformation temperature (Sandquist, 1987). In older low solids boilers the smelt temperatures are 750–810°C. In modern high solids boilers the typical smelt temperatures are 800–850°C. The smelt flow corresponds typically from 0.400 to 0.480 kg per kilogram of incoming black liquor dry solids flow.

11.2.2 Reduction and Sulfidity

The main process property of the smelt is the reduction. Reduction is the molar ratio of Na_2S to Na_2SO_4,

$$\text{Reduction} = \frac{Na_2S}{Na_2S + Na_2SO_4} \qquad (11.1)$$

The higher the reduction, the lower the amount of sodium that reaches the pulp mill cooking department as unusable. Reduction rates of 95–98% are not uncommon in well-operated recovery boilers. Usually the reduction efficiency increases as the char bed temperature increases. From thermodynamical equilibrium we can note that there should be a very small amount of sodium oxides and thiosulfate (Backman et al., 1996).

Sulfidity is the molar ratio of sodium sulfide to the total alkali content.

$$\text{Sulfidity} = \frac{S_{tot}}{Na_2 + K_2} \qquad (11.2)$$

This equation is widely in use because the ease of measuring sulfidity depends on the liquor circulation of the mill. Too high a sulfidity causes operating problems for the recovery boiler. In particular, increased sulfidity increases SO_2 and total reduced sulfur (TRS) emissions. Often the mill's analysis of the reduction rate is done for green liquor. Alkali in the green liquor will generally result in lower values that what is measured in smelt. Typically in modern mills the reduction in green liquor is 2–3% points lower than in smelt.

11.2.3 Sodium

Sodium is released during the black liquor burning and char bed reactions through vaporization and reduction of sodium carbonate. Sodium release increases as a function of temperature. At the beginning of combustion a large portion of sodium is connected to the organic portion of the black liquor. At the end of the volatiles release almost all of it is inorganically bound. Sodium release in kraft recovery boilers increases with an increasing lower furnace temperature. It has been assumed that in industrial boilers all of the electrostatic precipitator (ESP) dust is from reactions with vaporized sodium. In addition, the amount of sodium released as a function of carbonate in ESP dust seems to increase. An increase in carbonate indicates an increase in the lower furnace temperature (Vakkilainen, 2005). The sodium content in black liquor is around 20 w-%. This means that the sodium release in the recovery furnace is about 10% of the sodium in black liquor.

Much-studied reactions involving sodium are hydroxide formation, reduction reactions, sulfate formation with hydroxides, sulfate formation with chlorides, sulfate formation with carbonate, and carbonate formation.

11.3 Recovery Boiler Design

The recovery boiler has several purposes. The first is to convert all organic material in the black liquor to heat for high-pressure steam in an environmentally friendly way. The second is to recycle and regenerate spent chemicals in black liquor.

Figure 11.7 A modern recovery boiler at Santa Fe, Nacimento, Chile.

The third is to process several byproduct streams. There are two opposing processes happening at the same time. Concentrated black liquor is burned in the furnace. Burning requires oxidation. At the same time, reduced inorganic chemicals emerge molten from the smelt spouts. Reductive processes require the absence of oxygen (Fig. 11.7).

11.3.1 Key Recovery Boiler Design Alternatives

When designing recovery boilers, there are alternative solutions. Major recovery boiler design options are: screen or screenless superheater area design, single drum or two drums, lower furnace tubing material, furnace bottom tubing material, vertical or horizontal boiler bank and economizer arrangement, and the number and type of air levels.

11.3.2 Key Design Specifications

When sizing a recovery boiler, some key design specifications are usually given to the boiler vendor who is doing the design. Typical information that is given includes dry solids capacity (without ash), black liquor gross heat value (without ash), black liquor elementary analysis (without ash), black liquor dry solids percentage from evaporation (without ash), desired main steam conditions, feedwater inlet temperature, and economizer flue gas outlet temperature. Sometimes the desired superheated steam temperature control point—the percentage of maximum continuous rating(MCR)—is also given.

Black liquor dry solids flow is the key design criteria. It establishes the required size of the boiler. With elementary analysis and dry solids one can calculate the heat released in the furnace. With water and steam values the MCR steam flow is established. It should be noted that when black liquor is sprayed onto the furnace, it contains ash collected from the ESP and ash hoppers. Because ash-free black liquor is the input flow to the recovery plant, it is usually chosen as the design base.

11.3.3 Improving Air Systems

To achieve solid operation and low emissions the recovery boiler air system needs to be properly designed. Air system development continues and has been continuing as long as recovery boilers have existed (Vakkilainen, 2005). As soon as the target set for the air system has been met, other new targets are given. Currently the new air systems have achieved low NOx but are still working on lowering fouling (Table 11.1).

The first generation air system in the 1940s and 1950s consisted of a two-level arrangement: primary air for maintaining the reduction zone and secondary air below the liquor guns for final oxidation (Llinares and Chapman, 1989). The second generation air system targeted high reduction. In 1954 CE moved their secondary air from about 1 m below the liquor guns to about 2 m above them. The third generation air system was the three-level air system. At the same time, stationary firing gained ground. The use of about 50% secondary air seemed to give a hot and stable lower

Table 11.1 Development of Air Systems

Air System	Main Target	Additional Targets
First generation	Stable burning of black liquor	
Second generation	High reduction	Burn liquor
Third generation	Decrease sulfur emissions	Burn black liquor, high reduction
Fourth generation	Low NOx	Burn black liquor, high reduction, low sulfur emission
Fifth generation	Decrease superheater and boiler bank fouling	Burn black liquor, high reduction, low emissions

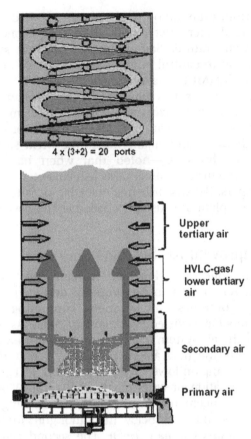

4 x (3+2) = 20 ports

Upper tertiary air

HVLC-gas/ lower tertiary air

Secondary air

Primary air

Figure 11.8 Principle of vertical air. From Kaila, J., Saviharju, K., 2003. Comparison of recovery boiler CFD modeling to actual operations. PAPTAC Annual meeting, January 28, Montreal, Canada, 13 p.

furnace. Higher black liquor solids (65–70%) started to be used. A hotter lower furnace and improved reduction were reported. With three-level air and higher dry solids the sulfur emissions could be kept in reasonably low level.

The fourth generation air systems are multilevel air and vertical air. As black liquor dry solids to the recovery boiler have increased, achieving low sulfur emissions is no longer the target of the air system. Instead, low NOx and low carryover are the new targets (Fig. 11.8).

11.3.4 Multilevel Air

The three-level air system was a significant improvement, but better results were required. Use of computational fluid dynamics (CFD) models offered a new insight of air system workings. The first to develop a new air system was Valmet (Tampella) with their 1990 multilevel secondary air in Kemi, Finland, which was later adapted for a string of large recovery boilers. They also patented the four-level air system, where an additional air level is added above the tertiary air level. This enables significant NOx reduction.

11.3.5 Vertical Air

Vertical air was invented by Uppstu (1995). His idea was to turn traditional vertical mixing into horizontal mixing. Closely spaced jets would form a flat plane. In traditional boilers this plane has been formed by secondary air. Placing the planes in a 2/3 or 3/4 arrangement improved mixing results. Vertical air has a potential to reduce NOx as staging air helps to decrease emissions (Forssén et al., 2000). In vertical air the primary air is arranged conventionally. The rest of the air ports are placed in an interlacing 2/3 or 3/4 arrangement.

11.3.6 Single Drum

All modern recovery boilers are of the single drum type. The single drum has replaced the two-drum (or bi-drum) construction in all but the smallest, low-pressure boilers. The same trend, but 20 years earlier, happened with coal-fired boilers.

11.3.7 Evolution of Recovery Boiler Design

There have been significant changes in kraft pulping in recent years. The increased use of new modified cooking methods and oxygen delignification has increased the degree of organic residue recovery. Black liquor properties have reflected these changes.

Changes in investment costs, increases in scale, demands placed on energy efficiency, and environmental requirements are the main factors directing the development of the recovery boiler (Vakkilainen, 2005). Steam generation increases with increasing black liquor dry solids content. For a rise in dry solids content from 65% to 80%, the main steam flow increases by about 7%. The increase is more than 2% per each 5% increase in dry solids. Steam generation efficiency improves slightly more than steam generation itself. This is mainly because the drier black liquor requires less preheating.

There are recovery boilers that burn liquor with solids concentration higher than 80%. Unreliable liquor handling, the need for pressurized storage, and high-pressure steam demand in the concentrator have frequently prevented sustained operation with very high solids. The main reason for the handling problems is the high viscosity of black liquor associated with high solids content. Black liquor heat treatment (LHT) can be used to reduce viscosity with high solids (Kiiskilä et al., 1993).

For pulp mills the significance of electricity generation from the recovery boiler has been secondary. The most important factor in the recovery boiler has been high availability. The electricity generation in recovery boiler process and steam cycle can be increased by elevated main steam pressure and temperature or by higher black liquor dry solids (Raukola et al., 2002).

Increasing the main steam outlet temperature increases the available enthalpy drop in the turbine. The normal recovery boiler main steam temperature of 480°C is lower than the typical main steam temperature of 540°C for coal- and oil-fired utility boilers. The main reason for choosing a lower steam temperature is to control superheater corrosion. The requirement for high availability and use of less expensive materials are often cited as other important reasons.

11.3.8 Modern Recovery Boilers

The modern recovery boiler is a single-drum design with a vertical steam-generating bank and wide-spaced superheaters. The most marked change around 1985 was the adoption of the single-drum construction. The construction of the vertical steam-generating bank is similar to that of the vertical economizer. The vertical boiler bank is easy to keep clean. The spacing between the superheater panels is increased and leveled off at over 300 mm but less than 400 mm. Wide spacing in superheaters helps to minimize fouling. This arrangement, in combination with sweetwater attemperators, ensures maximum protection against corrosion. There have been numerous improvements in recovery boiler materials to limit corrosion (Hänninen, 1994; Klarin, 1993).

The effect of increasing dry solids concentration has had a significant effect on the main operating variables. The steam flow increases with increasing black liquor dry solids content. Increasing the closure of the pulp mill means that less heat per unit of black liquor dry solids will be available in the furnace. The flue gas heat loss will decrease as the flue gas flow diminishes. Increasing black liquor dry solids is especially helpful since the recovery boiler capacity is often limited by the flue gas flow.

The nominal furnace loading has increased in the past 20 years and will continue to increase (McCann, 1991). Changes in air design have increased furnace temperatures (Adams et al., 1997; Lankinen et al., 1991, MacCallum, 1992; MacCallum and Blackwell, 1985). This has enabled a significant increase in hearth solids loading (HSL) with only a modest design increase in the hearth heat release rate (HHRR). The average flue gas flow decreases as less water vapor is present. Therefore the vertical flue gas velocities can be reduced, even with increasing temperatures in the lower furnace.

The most marked change has been the adoption of the single-drum construction. This change has been partly affected by more reliable water quality control. The advantages of a single drum boiler compared to a bi-drum are improved safety and availability. Single-drum boilers can be built to higher pressures and bigger capacities. Savings can be achieved with decreased erection time. There are fewer tube joints in the single-drum construction, so drums with improved startup curves can be built.

The construction of the vertical steam-generating bank is similar to the vertical economizer, which, based on experience, is very easy to keep clean (Tran, 1988). The vertical flue gas flow path improves the cleanability with a high dust loading (Vakkilainen and Niemitalo, 1994). To minimize the risk of plugging and to maximize the efficiency of cleaning, both the generating bank and the economizers are arranged with generous side spacing (Mäntyniemi and Haaga 2001). The two-drum boiler bank blockage is often caused by too tight spacing between the tubes.

The spacing between the superheater panels has increased. All superheaters are now wide-spaced to minimize fouling. This arrangement, in combination with sweetwater attemperators, ensures maximum protection against corrosion. With wide spacing, plugging of the superheaters becomes less likely, the deposit cleaning is easier, and the sootblowing steam consumption is lower. An increased number of superheaters facilitates the control of superheater outlet steam temperature, especially during startups.

The lower loops of the hottest superheaters can be made of austenitic material, with better corrosion resistance. The steam velocity in the hottest superheater tubes is high, decreasing the tube surface temperature. Low tube surface temperatures are essential to prevent superheater corrosion. A high steam side pressure loss over the hot superheaters ensures uniform steam flow in the tube elements.

11.3.9   State of the Art and Current Trends

Recovery boiler design changes slowly. There are, however, some features that boilers bought today have in common. A state-of-the-art recovery boiler has the following features:
- One-drum boiler with three-part superheater and water screen (optional)
- Steam design data at least 9.2 MPa/490°C
- Design as fired black liquor dry solids over 80% with pressurized heavy liquor storage tank
- Liquor temperature control with flash tank, indirect liquor heaters for backup
- Dilute noncondensable gases (DNCG) burning in the boiler
- Low emissions of TRS, SO_2, and particulates
- Flue gas cleaning with ESP (no scrubbers)

The design changes occurring can be listed. Current trends for recovery boilers are:
- Higher design pressure and temperature due to increasing demands of power generation
- Use of utility boiler methods to increase steam generation
- Superheater materials of high-grade alloys
- Further increase in black liquor solids toward 90% by concentrators using elevated steam pressure
- Burning of biological effluent treatment sludge and bark press filtrate effluent
- Concentrated noncondensable gases (CNCG) burner (low volume, high concentration (LVHC) gases)
- Dissolving tank vent gases returned to the boiler
- Advanced air systems for NOx control.

11.4   Heat Transfer Surface Design and Material Selection

When recovery boilers are designed, one of the most difficult questions that arises is what kind of materials should be used for different parts of the boiler. Corrosion is typically divided into areas based on the location of the corrosion: water side corrosion, high-temperature corrosion, and low-temperature corrosion.

Water side corrosion occurs in the steam–water side of the boiler tubes. Usually the cause is impurities in the feedwater. High-temperature corrosion typically occurs in the superheaters. Low-temperature corrosion occurs in the economizers and air heaters. Low-temperature corrosion is often associated with the formation of acidic deposits.

11.4.1   Furnace Design and Materials

Recovery boiler furnace walls and floors have long been under investigation to find better materials. In particular the floor construction and materials affect the recovery boiler safety (Bauer and Sharp, 1991). Most of the critical leaks in the furnace occur in the lowest 3 m of the furnace walls. Traditional protection for the lowest part is studding and refractory. Corrosion protection with studs is excellent, but this solution requires a large amount of maintenance and repair work. A membrane wall with welded corrosion protection of alloyed material is another corrosion protection option . A welded furnace wall is of comparable price to compound tubing, which is the most used recovery boiler wall construction. All new recovery boilers are of membrane design. Tangent tubing was phased out in the late 1980s (Sandquist, 1987).

It is important to protect the floor tubes from high temperatures. The proper design of water circulation lowers the maximum temperatures. Sufficient water flow needs to be maintained in the tubes to cool them and to remove steam bubbles that have been created. Usually the requirement for flow velocities is ≥ 0.5 m/s in all tubes.

The floor angle in modern boilers needs to be upward with the flow. As the bottom tubes are supported by steel beams, they hang a little. The floor angle helps to avoid parts where the steam bubbles could get stuck. Depending on the distance between the support tubes, the angle needs to be from 2.5 to 4 degrees. Smelt spouts need to be high enough so that all the floor is covered with a frozen smelt layer. Especially critical is the area farthest from the smelt spouts and the area right in front of the smelt spouts. In practice it seems that 200–300 mm is enough. Too much height will cause problems when one tries to empty the bed for shutdown.

11.4.2   Furnace Tube Materials

Some of the most typical furnace tube materials are listed in Table 11.2. Many more have been tried and for one reason or another abandoned. Carbon steel was the material of choice before the compound tubing. The upper furnace above the highest air level is always made from carbon steel. Carbon steel seems to resist most corrosive conditions at oxygen-rich conditions. Carbon steel has also been used recently as floor material,

Table 11.2 Properties of Typical Floor Tube Materials

	Carbon Steel	304 L	Sanicro 38 (Alloy 825)	Sanicro 65 (Alloy 625)
Main elements	Fe	20Cr—10Ni	20Cr—40Ni	20Cr—60Ni
Thermal expansion, $10^{-6}/°C$	13.5	17.5	14.9	13.9
Thermal conductivity, W/m°C	41	19	16	14
Stress corrosion cracking resistance	Excellent	Low	High	Excellent
Corrosion resistance	Low	Moderate	Excellent	High

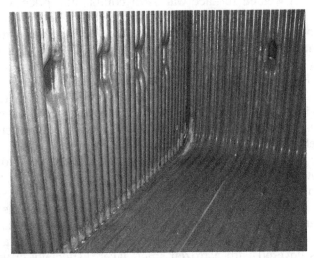

Figure 11.9 Modern carbon steel furnace bottom. Courtesy of Andritz.

Figure 11.9. Floors with carbon tube are not susceptible to stress corrosion cracking (SCC) corrosion. It should be noted that carbon tubes cannot resist firing black liquor on bare tubes nor smelt. Some care should be taken when operating recovery boilers with carbon steel floors.

Extensive research in Finland has been carried out in which the corrosion of different materials in molten polysulfides was studied. This research showed that Sanicro 38-type composite material had the best corrosion resistance among the steels studied (Mäkipää and Backman, 1998). Test panels made of Sanicro 38 installed in 1991 and 1994 have not shown any alarming corrosion; nor has there been any reported cracking in recovery boiler bottoms made from Sanicro 38 since 1995. This highly alloyed material seems to have good corrosion resistance but it is fairly expensive.

Stainless steel 304 L seems to last well in the furnace walls above the char bed. It is very resistant to sulfidation. SCC in the tubes at the furnace bottom has made manufacturers and recovery boiler owners search for replacement materials in that area (Keiser et al., 2004).

Suppliers' current recommendations are to use modified alloys in the front and rear bends and close to the side walls. To facilitate weld inspection the whole lower furnace is often made of modified alloys up to and over the primary air ports. The present favorite is Sanicro 38 composite tubing. In addition to the high content of chrome and nickel, the tube has about the same thermal expansion coefficient as the carbon steel.

Sanicro 65 (Alloy 625) composite tubing is another possibility. It has very favorable properties with regard to thermal fatigue and SCC. There are some reports of failure. Therefore the use of 625 needs further research at the present time. Another area under study is air port cracking (Keiser et al., 2002). Primary air ports and smelt openings seem to exhibit cracking. Thermal cycling and smelt contact are suspected causes.

The use of compound tubing in recovery boilers dates back to the 1980s. Compound tubing is expensive and the selection of materials is limited. A competing alternative is the chromizing of tubes (Plumley et al., 1989; Labossiere and Henry, 1999). Another popular method is spraying on a plasma coating. The compound surface can also be replaced by a welded surface. Of all the above methods, quality control and inspection is easiest with compound tube.

11.4.3 Superheater Design and Materials

Recovery boilers suffer from superheater corrosion. Corrosion is the main problem that limits the ability of the kraft recovery boiler to produce electricity (Bruno, 2003). In coal-fired boilers much higher superheater temperatures are typically used. In comparison to coal-fired boilers, kraft recovery boilers have higher rates of alkali metals, chloride in gaseous form and often highly reducing conditions caused by carryover particles. On the other hand, the contents of some high-temperature corrosion-causing substances such as antimony, vanadium, and zinc are typically low.

Loss of tube thickness can be caused by sulfidation and alkali or chloride corrosion. Typically superheaters exhibit higher corrosion resistance if their tube materials have a higher content of chromium (Blough et al., 1998).

11.4.4 Effect of the Steam Outlet Temperature

The main steam temperature is the main parameter that affects the choice of superheater materials. The rule of thumb is to keep the superheater surface temperature below the first melting temperature of the deposits (Salmenoja and Tuiremo, 2001). Corrosion rates in the final superheaters are increased because the superheater material temperatures are high. As can be seen, there is typically some temperature range where the corrosion rate is acceptable. Increasing tube temperature by some tens of degrees can significantly increase the corrosion rate.

Steam side heat transfer coefficients in typical recovery boiler superheaters are low. The superheater surface temperature can be tens of degrees higher than the bulk steam temperature. It can easily be seen that surface temperatures and thus corrosion rates are greatly affected by superheater positioning. Furnace radiation can effectively be reduced by placing a screen to block the radiation heat flux. Therefore placing the hottest superheaters behind the nose or screen will significantly decrease corrosion.

11.4.5 Typical Superheater Materials

A typical primary material for superheaters, when they are protected from direct furnace radiation, is carbon steel. Secondary and tertiary superheater materials often contain 1–3% Cr. These kinds of material are easy to weld and have good corrosion protection. T22/10CrMo910 material can usually be used up to 495°C steam outlet temperature (Clement, 1990). With higher temperatures and higher chloride and potassium content in the black liquor it is advisable to use higher chromium-containing tubes.

Fujisaki et al. (1994) found that recovery boiler superheater corrosion is much reduced when the chrome content of the superheater tube is increased. A similar trend was found in Swedish studies in the Norrsundet recovery boiler (Eriksson and Falk, 1999). The researchers found that alloyed austenitic materials 304 L and Sanicro 28 had much better corrosion resistance than highly alloyed ferritic materials SS2216 and X20. Stainless steel lower bends in the hottest superheaters have been used for decades.

11.4.6 Boiler Bank Design and Materials

Two-drum boiler banks in recovery boilers suffer from mud drum corrosion (Labossiere and Henry, 1999). This type of corrosion is caused by steam from sootblowing wetting the salt at the tube joints in the lower drum. The progress of the near drum corrosion can be monitored with ultrasonic equipment (Soar et al., 1994).

One problematic type of failure is caused by vibrations from sootblowing. The longer the free tube length, the higher the resulting stress at the joins. Industry practice states that the maximum length of free tubes is some 8 m. Typically longer tubes are too flexible and will vibrate too much. This will create cracks and faults after a few years. A finned design causes temperature differences between the fin and the tube. This will create high stresses at the fin ends. To prevent these stresses cut fins are preferred.

Some blockage problems have been reported on the lower end of boiler banks (Sandquist, 1987). If the lower headers are located too close to each other, they trap falling material. Placing a sootblower close to the lower end is also critical.

11.4.7 Economizer Design and Materials

Modern economizers are of vertical design. The earliest horizontal economizers had severe plugging problems and were replaced by a cross-flow design. The cross-flow economizer had lower heat transfer coefficients and was more prone to plugging than the modern vertical economizer. In economizers the loss of tube thickness can be caused by gas side corrosion, sulfidation and acid dew point corrosion, or water side erosion corrosion. The lower ends of economizers in recovery boilers suffer from water side erosion corrosion. Typically the symptoms are worst in the first few meters of the economizer tube.

Recovery boiler economizers have hundreds of weld joints. Each weld, even after inspection, is potentially problematic. Therefore the preference has been to avoid unnecessary welds and use only continuous tubes without butt welds. The largest boilers have economizer lengths of 27 m. The maximum length of carbon steel tubes is some 23 m. Therefore in the newest boilers this preference for continuous tubes cannot be adhered to. Attention should be paid to the quality of welds in economizer tube joints.

References

Adams, T.N., Frederick, W.M., James, G., Thomas, M., Hupa, M., Iisa, K., et al., 1997. Kraft recovery Boilers. AF&PA, TAPPI press, Atlanta, 381 p. ISBN 0962598593.

Backman, R., Skrifvars, B.-J., Hupa, M., Siiskonen, P., Mäntyniemi, J., 1996. Fluegas and dust chemistry in recovery boilers with high levels of chlorine and potassium. J. Pulp Paper Sci. 22 (4), J119–J126.

Bauer, D.G., Sharp, W.B.A., 1991. The inspection of recovery boilers to detect factors that cause critical leaks. TAPPI J. 74 (9), 92–100.

Blough, J.L., Seitz, W.W., Girshik, A., 1998. Fireside Corrosion Testing of Candidate Superheater Tube Alloys, Coatings, and Claddings - Phase 2 Field Testing. Technical Report, ORNL/93-SM401/02. Oak Ridge National Lab, TN, 75 p.

Bruno, F., 2003. Corrosion as a cause for recovery boiler damages. Pulp Pap Can. 104 (6), 143–151.

Clement, J.L., 1990. High pressure and temperature recovery boilers. Proceedings of Babcock & Wilcox, Pulp & Paper Seminar. February 12–14, 1990, Portland, Oregon, 7 p.

Eriksson, T., Falk, I., 1999. Överhettarmaterial för energi-effektivare, miljövänligare och bränsleflexiblare sodahus och barkpannor (Superheater materials for better energy efficiency, lower emissions and better fuel flexibility in recovery and bark boilers). Studsvik AB, S6-615, Värmeforsk Service AB. 47 p. (in Swedish).

Forssén, M., Kilpinen, P., Hupa, M., 2000. NOx reduction in black liquor combustion—reaction mechanisms reveal novel operational strategy options. TAPPI J. 83 (6), 13 p.

Fujisaki, A., Tateishi, M., Baba, Y., Arakawa, Y., 2003. Plugging prevention of recovery boiler by character improvement of the ash which used potassium removal equipment (Part II). Pulp Pap. Can. 104 (1), T1–T3.

Fujisaki, H., Takatsuka, M., Yamamura, 1994. World's Largest high pressure and temperature recovery boiler. Pulp Pap. Can. 95 (11), T452.

Holmlund, K., Parviainen, K., 2000. Evaporation of black liquor. In: Gullichsen, J., Fogelholm, C.-J., (Series Eds.), Chapter 12 in Chemical Pulping, Book 6, Finnish Paper Engineers' Association and TAPPI. ISBN 9525216063.

Hänninen, H., 1994. Cracking and corrosion problems in black liquor recovery boilers. 30 Years Recovery Boiler Co-operation in Finland. International conference, Baltic Sea, May 24–26, pp. 121–132.

Kaila, J., Saviharju, K., 2003. Comparison of recovery boiler CFD modeling to actual operations. PAPTAC Annual meeting, January 28, Montreal, Canada, 13 p.

Kelly, P.A., Frederick, W.J., Grace, T.M., 1981. The residence time distribution of inorganic salts in kraft recovery boilers. TAPPI J. 64 (10), 85–87.

Keiser, J., Singbeil, D.L., Sarma, G.B., Choudhury, K., Singh, P.M., Hubbard, C.R., et al., 2002. Comparison of cracking in recovery boiler composite floor and primary air port tubes. TAPPI J. 1 (2), 1–7.

Keiser, J.R,, Singbeil, D.L., Sarma, G.B., Kish, J.R., Choudhury, K.A., Frederick, L.A., et al. 2004. Cracking and corrosion of composite tubes in black liquor recovery boilers. 40th Anniversary International Recovery Boiler Conference, Finnish Recovery Boiler Committee. Haikko Manor, Porvoo, May 12–14, 2004, pp. 59–89.

Kiiskilä, E., Lääveri, A., Nikkanen, S., Vakkilainen, E., 1993. Possibilities for new black liquor processes in the pulping industry energy and emissions. Bioresour. Technol. 46, 129–134.

Klarin, A., 1993. Floor tube corrosion in recovery boilers. TAPPI J. 76 (12), 183–188.

Labossiere, J., Henry, J., 1999. Chromizing for near-drum corrosion protection. TAPPI J. 82 (9), 150–157.

Lankinen, M., Paldy, I.V., Ryham, R., Simonen, L., 1991. Optimal solids recovery. Proceedings of CPPA 77th Annual meeting, pp. A373–A378.

Llinares, Jr. V., Chapman, P.J., 1989. Stationary firing, three level air system retrofit experience. Proceedings of 1989 TAPPI Engineering Conference. Atlanta, Georgia, September 10–13, 10 p.

MacCallum, C., 1992. Towards a superior recovery boiler air system. Proccedings of 1992 International Chemical Recovery Conference. Seattle, Washington, June 7–11, pp. 45–56.

MacCallum, C., Blackwell, B.R., 1985. Modern kraft recovery boiler liquor-spray and air systems. Proceedings of 1985 International Chemical Recovery Conference, New Orleans, LA, pp. 33–47.

McCann, C., 1991. A review of recovery boilers process design. CPPA 77th Annual Meeting, pp. A49–A58.

McCarthy, J.H., 1968. Recovery plant design and maintenance. In: Whitney, R.P. (Ed.), Chapter 5 of Chemical Recovery in Alkaline Pulping Process, TAPPI Monograph Series. No. 32, Mack Printing Company, Easton, PA, pp. 159–199.

Mäkipää, M., Backman, R., 1998. Corrosion of floor tubes in reduced kraft smelts: studies on effects of chlorine and potassium. Proceedings of 9th International Symposium on Corrosion in the Pulp and Paper Industry. May 26–29, 1998, Ottawa, Ontario, Canada.

Mäntyniemi, J., Haaga, K., 2001. Operating experience of XL-sized recovery boilers. Proceedings of 2001 TAPPI Engineering, Finishing & Converting Conference. TAPPI Press, Atlanta, GA, 7 p.

Plumley, A.L., Lewis, E.C., Barker, T.J., Esser, F.A., 1989. Chromizing for recovery boiler corrosion protection. Proceedings of 1989 TAPPI Engineering Conference. Atlanta, Georgia, September 10–13, 8 p.

Raukola, A.T., Ruohola, T., Hakulinen, A., 2002. Increasing power generation with black liquor recovery boiler. Proceedings of 2002 TAPPI Fall Technical Conference. September 8–11, San Diego, CA, 11 p.

Salmenoja, K., Tuiremo, J., 2001. Achievements in the control of superheater corrosion in black liquor recovery boilers. Proceedings of 2001 TAPPI Engineering, Finishing & Converting Conference. TAPPI Press, Atlanta, GA, 9 p.

Sandquist, K., 1987. Operational experience of composite tubing in recovery boiler furnaces. Götaverken Energy Systems, Technical Paper TP-2-87, 15 p.

Saviharju, K., Pynnönen, P., 2003. Soodakattilan keon hallinta (Control of Kraft recovery boiler char bed). Konemestaripäivä, 23.1.2003, Oulu, Finnish recovery boiler users association, 16 p. (in Finnish).

Soar, R.J., Bardutz, R.W., Guzi, C.E., 1994. Characterizing the wastage on near-drum generator tubes using full-coverage ultrasonic scans. TAPPI J. 77 (8), 201–209.

Tran, H., 1988. How does a kraft recovery boiler become plugged. TAPPI 1988 Kraft Recovery Operations Seminar, pp. 175–183.

Uppstu, E., 1995. Soodakattilan ilmanjaon hallinta (Control of recovery boiler air distribution). Soodakattilapäivä 1995, Finish recovery boiler committee, 6 p (in Finnish).

Vakkilainen, E.K., 2000. Estimation of elemental composition from proximate analysis of black liquor. Paperi ja Puu-Paper and Timber. 82 (7), 450–454.

Vakkilainen, E.K., 2005. Kraft recovery boilers—principles and practice. Suomen Soodakattilayhdistys r.y., Valopaino Oy, Helsinki, Finland, 246 p. ISBN 9529186037.

Vakkilainen, E., Ahtila, P., 2011. Modern method to determine recovery boiler efficiency. O Papel. 72 (12), 58–65.

Vakkilainen, E., Niemitalo, H., 1994. Measurement of high dry solids fouling and improvement of sootblowing control. Proceedings of 1994 TAPPI Engineering Conference, San Francisco, California.

12

MEASUREMENTS AND CONTROL

No modern boiler can operate without proper automation and control. Although in the early days boiler operation employed dozens of men, now often one sees only a few and those are sitting in a control room. It has been recognized that energy efficient operation requires good control. One has to continuously balance air and fuel. Simultaneously one must tune the boiler to minimize emissions. All this requires sophisticated controls.

The ability to exercise control requires proper measurements. Temperatures, pressures, and flows must be measured and recorded.

Steam generators are pressure devices and can blow up (Effenberger, 2000). The safety of workers and equipment demands interlocks. Interlocks are safety devices that are used to prevent accidents caused by improper operation. Interlocks prevent operators from making fatal mistakes.

Steam Generation from Biomass. DOI: http://dx.doi.org/10.1016/B978-0-12-804389-9.00012-5

12.1   Interlocks

wAll interlocks used to be hard-wired. In modern automation sometimes programmable logic circuits are used. The main function of interlocks is safety and minimizing damage to property. There are basically two types of interlock: event forcing and sequence forcing. Legislation, codes, and insurers have specifications for minimum requirements for interlocks in boiler plants.

Typical event interlocks are:
- Drum minimum level -> stops fuel firing.
- Drum maximum level -> stops fuel firing.
- Loss of air fan -> stops fuel firing.
- Loss of induced draft (ID fan) -> stops fuel firing.
- Loss of flame -> stops fuel firing.

Typical sequence interlocks ensure that:
- Firing cannot be started unless the fan is running.
- The main fuel cannot be started if the pilot flame is out.
- The flue gas damper is closed when the fan stops.
- The fuel main valve is closed when the flame is out.
- Open gas is purged when the flame is out.

The sequence interlock often has timed delays. In practice this means, for e.g., that the furnace must be purged 30 seconds before the fuel can be introduced.

12.2   Boiler Control Principles

In a modern boiler there are about a thousand control loops. Valve positions, fan and pump operation, as well as various levels, are controlled in a similar way to other processes. However, there are only a few main boiler controls. These are load control, drum level control, fuel rate control, air flow control, superheating control, and furnace draft (Lindsley, 2000).

12.2.1   Steam Flow Control

Load or steam flow is controlled by burning the desired amount of fuel, Fig. 12.1. Load control has typically 10–20 minutes time lag due to the thermal capacity of the pressure vessel.

The most problematic aspect of load control is trying to ascertain that a desired fuel flow is introduced to the boiler. This is easy for oil and gas, which are relatively homogenous fuels, but difficult for other fuels for the following reasons:
- In fluidized bed, circulating fluidized bed (CFB), grate, and recovery boilers there can be a significant amount of

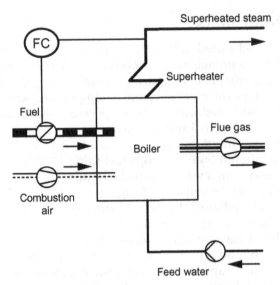

Figure 12.1 Load control. *FC*, flow control.

unburned fuel remaining in the furnace, which will react even if the fuel flow is changed.
- Nonhomogenous fuel flows (peat, bark, and waste) are difficult to measure.
- Many fuel injecting devices (belts, screws) tend to produce an uneven fuel flow.

12.2.2   Drum Level Control

One of the most challenging applications in modern boilers is the drum level control, Fig. 12.2. This is because there are significant time lags involved. Using separate loops to tune the same things often makes the control shaky.

The drum level control functions are as follows:
1. The feedwater flow is measured.
2. The feedwater flow and steam flow are used to set the desired flow.
3. The flow is controlled by the valve position (fast control).
4. The proper pressure difference is maintained by the pump speed control (slow control).
5. Changes in the drum level are measured.
6. The feedwater flow is derived from the steam flow, load change, and the change in the drum level.
7. The load change information is used to indicate the desired steam flow.

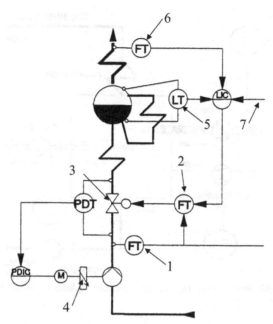

Figure 12.2 Functioning of drum level control. *FT,* flow transmitter; *LT,* level transmitter; *PDT,* pressure difference transmitter; *M,* motor; *LIC,* level indicator and control; *PDIC,* pressure difference indicator and control.

12.2.3   Fuel Rate Control

The fuel rate control is in most cases a simple feedback cycle where the fuel flow is first measured. The control circuit receives it and compares it to the desired rate. Rate change information is sent to the fuel rate control device. For liquid fuels this is usually a valve and sometimes a pump (Saviharju and Pynnönen, 2003). For solid fuels this is a screw, belt, or feeder. With solid fuels, measurement of the true fuel flow is difficult. Therefore the steam flow is used in connection with, e.g., screw rotation to arrive at the fuel flow rate (Fig. 12.3).

12.2.4   Air Flow Control

Since the introduction of reliable O_2 measurements in the 1980s, the air flow control has been based on maintaining the desired O_2 level in the flue gases. Primary air flow control is done by maintaining a constant air/fuel ratio. This means that when the load control requires a change in the fuel flow, then the air flow control makes a proportional change in the air flow. The secondary air flow control adjusts the flow to match the O_2

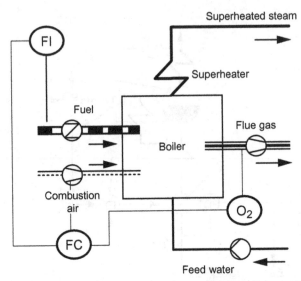

Figure 12.3 Air flow control. O_2, oxygen indicator; *FI*, flow indicator; *FC*, flow control.

measurement in the stack. This means that the user sets an O_2 target and the control fine tunes the secondary air flow to reach this target (Kortela and Marttinen, 1985). If the boiler has tertiary or quaternary air, then they are often adjusted in a similar to the primary air flow (i.e., based on a desired percentage of total air flow).

In the past the only air flow control method was changing the position of the dampers. They are still used to hand tune the introduction of air. Their drawback is the pressure losses causing high own electricity consumption. The cheapest air fan control is fitted with guide vanes, which change the inlet air angle. The inlet vane has high efficiency at the nominal operating load. Efficiency drops at partial loads. The fan speed can be controlled by hydraulic couplings or an inverter. Both are reliable methods. Hydraulic couplings are cheaper, but have higher losses. Inverters are currently the best method to save energy.

12.2.5 Furnace Draft

Flames and flue gas must be kept in the furnace enclosure where the fuel burning takes place. This requires operating at an underpressure of 300–1000 Pa. Because of underpressure, the flow in all openings is from the boiler room to the furnace. This underpressure was formerly caused by the draft in the chimney from which the name draft control is derived. Draft is normally regulated by induced fans. Furnace pressure is measured at the top of the furnace and this information is used to control the fans. There is typically ash depositing, which blocks these measurements. Frequent hand opening by the operator is required to keep the fans running.

12.2.6 Superheating Control

Usually we want the boiler to have a wide operating range: e.g., 60–100%, where we get the full superheat. Typically the amount of steam temperature increase gets larger with the load increase, Fig. 12.4. Thus from a certain load the steam temperature is higher than desired. We must therefore control the superheater temperature to the desired level. Decreasing the main steam temperature is called desuperheating.

We note that at 60% load there is no need for any desuperheating. At 100% load we need the highest amount of desuperheating. Because of the part load superheating requirement, we must build more of a superheating surface than the minimum surface required for full superheating at full load (Singer, 1991).

Depending on the place of desuperheating, at partial load the material temperature at the primary or secondary superheater can be, in some applications, higher than at full load. When we design superheaters, we must look at all loads to determine the maximum allowable operating temperature and pressure for each heat transfer surface (Fig. 12.5).

The most typical method of desuperheating is to spray feedwater, which evaporates and cools the steam. Feedwater spraying requires strict control of feedwater quality as impurities will cause plugging and corrosion in superheaters and the turbogenerator blades.

When feedwater is sprayed onto the steam line, it is atomized to fine droplets. The steam–water mixture goes toward equilibrium as the temperature of the steam decreases and the

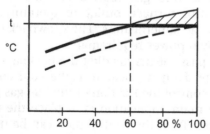

Figure 12.4 Superheating control.

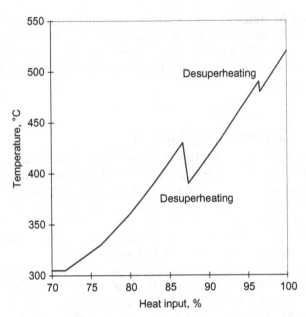

Figure 12.5 Desuperheating.

temperature of the water droplets increases. When the water droplets reach a temperature corresponding to the saturated temperature at steam pressure, the water droplets start to evaporate. This cools the steam even more. Typically the evaporation is completed and the whole mixture is steam within a few meters of the spray connection. Feedwater spraying is the most common method of controlling steam temperature because it is simple and cheap.

Especially in large CFB boilers the flue gas recycle affects superheating. Flue gas to be recycled is drawn from the backpass end before the air preheater and blown with air to the lower furnace. With more flue gas the furnace exit temperature decreases as does the radiative superheating, but the convective superheating increases because of the higher flue gas velocity. The flue gas recycle is used to control NOx formation, control furnace outlet temperature, and control ash slagging in the superheaters. The drawback of the flue gas recycle is the own power requirement.

The control pass means dividing the backpass into two parallel passes. The damper position at the cold end of the pass can be used to control the fraction of the flue gas going through each pass. The main application is to place the reheating surface on one pass. Then the control pass can be used to control the ratio of superheating to reheating.

The oldest method of controlling superheating is to regulate the flue gas flow by blowing less or more air into the furnace. Economically this is not wise as the increased O_2 content in the flue gas increases the flue gas losses.

12.2.7 Dolezahl or Steam Condensate Desuperheat

When the feedwater quality is low or the boiler uses lots of process condensate, it is inadvisable to use feedwater spraying. Then the water for spraying is condensed in a separate heat exchanger from the saturated steam with the feedwater. Usually the feedwater that has passed though the economizers is used.

The drawback of the Dolezahl system is that often the pressure differences available for the spraying are quite low. Therefore this limits the available desuperheating capacity.

12.3 Benefits of Process Control

The benefits of modern controls are more production, higher-quality output, savings in own energy use, improved safety, and increased environmental benefits. The drawbacks of modern control are that it is expensive, it requires training (computer skills), process know-how is lost, and more electrical maintenance is required. In addition, applying modern controls reduces the required workforce. The challenges of the process control are process simulation, forecasting, and especially measurement and process data presentation.

Process can be modeled to predict the correct intermediate values. Process simulations can be used in process optimization, training, and fault simulation. For example, nuclear power plants have been thoroughly tested by fault simulators. The main drawback is the lack of process models and lack of processing power.

Many plants create overviews and data analysis from those available at digital control system (DCS) with management information systems (MIS). Production information, fuel consumption logging, and power purchases can be tracked. With modern systems the process can be looked at more accurately. Missing and malfunctioning measurements can be seen immediately.

The role of forecasting is to present a viable scenario of the future. If we can estimate future values, money can be saved and the process can be better tuned. A typical example is electricity production in the next 24 hours. Forecasting is always difficult (otherwise the stock markets would go bust).

With optimizing systems we seek the best possible process operation. This means that quality, output, etc. can be maximized. Optimization of production from several power plants is a typical application. The optimum is often difficult to define and calculate mathematically.

In control rooms there are dedicated screens or printers that frequently churn out reports on faults. Fault detection is used to identify problematic process components and measurements. A typical error message could be: "Process shutdown from fault in motor ADS0123GF01." Fault signals do not always point to the "real" fault. Often single faults can create dozens of errors.

12.4 Measurements

Boiler operation requires reliable measurement of pressure, temperature, fluid flow, and steam quality. Firing requires additional safety devices, such as flame monitors.

12.4.1 Mandatory Measurements

There are often mandatory measurements. Typically laws require boilers to have several operating, local measurements such as drum level, operating pressure, and operating temperature (Stultz and Kitto, 1992).

Steam pressure needs to be controlled so that the operation of the boiler plant is kept under control. The boiler plant mechanical design sets a maximum allowable working pressure. If this level is exceeded, safety valves are opened to relieve pressure.

Steam temperature exiting the boiler needs to be tightly controlled. Temperature sets metallurgical constraints on the main steam line and the superheater tubes. They last longer if the steam temperature excursions are kept at a minimum. The turbine delivers lower electrical output if we lower the steam temperature.

The drum level indicator, Fig. 12.6, is typically a quartz tube between two valves. This type of device has been in use since the 1830s. The water level can be seen in the tube. Optical devices can be used to improve visibility. Output from the drum level indicator can be transferred to the DCS. The actual water level measurement in the drum is usually done with a pressure difference measurement. One tube connects to the lower part of the drum, another to the upper part of the drum. The water height can be calculated from the pressure difference, using known water and steam densities.

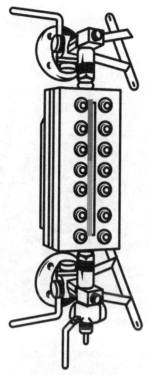

Figure 12.6 Drum level indicator.

12.4.2   Local Measurements

Local pressure measurements at the drum, after the feed-water pump and at the main steam outlet are the main local pressure measurements used. Measurement that directly shows the value for an operator standing nearby is called local measurement. Laws governing the safety of the boiler still require these. Remote measurements are often doubled by local measurements, Fig. 12.7. Process variables occasionally also need to be displayed locally. Local operation can also be of interest. For these, local displays are used.

12.4.3   Remote Measurements

Remote measurements actually have their measurements done locally. The result is transmitted to the DCS electrically. Placement is crucial for reliable operation, Fig. 12.8. Remote measurement consists of the actual measurement, conversion

Figure 12.7 Local pressure measurement.

Figure 12.8 Remote temperature measurement.

of measurement to suitable signal, typically done in a box resembling a can and the cable.

12.4.4 Typical Measurements

Temperature measurement is normally done by PT100 or thermocouple measurements. Either a current is generated or resistance is measured. An actual temperature measuring sensor is often inserted into the desired location inside a protective tube.

Pressure is frequently transmitted through a small tube to a sensor. Inside the sensor there is usually a membrane, which generates the pressure signal. Pressure changes alter the membrane dimensions, which change the electrical signals through it. The signal is transformed and sent to the control.

The mass flow measurement of fluids is often done using a magnetic flowmeter. Another alternative is to measure the time it takes sound waves to go through a flowing liquid. The traditional alternative is to use pressure difference over a known orifice or venturi. Almost always the mass flow measurement measures the volumetric flow, which needs to be transformed to mass flow though density measurement, function, or a table.

Mass flow measurement of solid flows (e.g., biomass fuel flow) is often rather cumbersome. In practice the speed of the belt, screw, or rotary feeder rotation is often used as an indicator of fuel flow.

The flame detector is used to ascertain the existence of a flame. Simple flame detectors use a change of conductivity or total irradiation as their signal. Far better monitors use a specific radiation band as a signal for flames. With this method, fouling, partial blocking of view and hot surfaces cannot cause a misleading signal.

In multiburner boilers the flame detector placing must be given special attention. It is easy for one flame detector to "see" the flame of an adjacent burner.

In modern installations oxygen is analyzed directly and continuously in the flue gas. Typically the device measures oxygen in wet gas. Due to the gradual loss of function caused by impurities, the sensors last 8−36 months.

References

Effenberger, H., 2000. Dampferzeugung (Steam boilers). Springer Verlag, Berlin, p. 852. ISBN 3540641750 (in German).

Kortela, U., Marttinen, A., 1985. Modelling, identification and control of a grate boiler. Proceedings of the American Control Conference, pp. 544−549.

Lindsley, D., 2000, Power-plant control and instrumentation: the control of boilers and HRSG systems. Institution of Electrical Engineers, London, 222 p. ISBN 0852967659.

Saviharju, K., Pynnönen, P., 2003. Soodakattilan keon hallinta (Control of Kraft recovery boiler char bed). Konemestaripäivä, 23.1.2003, Oulu, Finnish Recovery Boiler Users Association, 16 p. (in Finnish).

Singer, J.G., (Ed.), 1991. Combustion Fossil Power, fourth ed. Asea Brown Boveri, 977 p. ISBN 0960597409.

Stultz, S.C., Kitto, J.B., (Eds.), 1992. Steam Its Generation and Use, 40th ed. 929 p. ISBN 0963457004.

BOILER MANUFACTURE, ERECTION, AND MAINTENANCE

Boiler mechanical design deals with the design of boiler parts. In addition to the design of pressure parts we must design the mechanical structure. Boiler structures have changed a lot over the years. However, many old designs are still used today.

All large boilers are hung from top beams because of thermal stresses. Boiler buildings are nowadays almost always of steel structure. Working access to the boiler plant is with platforms. The pressure part is insulated to decrease the heat losses.

13.1   Steel Support

Large boiler buildings are mostly of steel structure (Fig. 13.1). Platforms are made from grating. Cross-beams transfer outside loads (wind, earthquake). Concrete is the next popular alternative to steel.

Steam Generation from Biomass. DOI: http://dx.doi.org/10.1016/B978-0-12-804389-9.00013-7

Figure 13.1 Boiler steel structures being erected.

13.1.1 Hanging Boilers

The pressure vessel hangs from the top of the steel structure. The vertical load is supported by the main steel. Usually 4–8 large vertical beams form the main support. Horizontal and cross-beams help to keep these beams standing up.

Topmost are the large main beams from which the boiler hangs. The main beams transfer this load to the main support beams. The height of the main support beams is from 300 to 3000 mm and the length from 10 to 30 m.

It would be difficult to connect all hanging weights directly to the main beams. Therefore a net of smaller secondary and tertiary beams is built on top of the primary beams. From these hang steel rods, Fig. 13.2.

Steel rods are used to transfer the weight of the pressure part to the supporting beams. Rods are ideal as they transfer only a one-dimensional load. Rods can be lengthened and shortened easily to balance the boiler weight. The rods also allow the thermal expansion of the pressure part to take place (Fig. 13.3).

Pressure in the furnace creates forces on the walls. Typical design pressures range from ± 5000 to $\pm 10,000$ Pa. High negative pressure is caused when air or fuel is switched off and flue gas fans are running. High positive pressure is caused when, e.g., a flue gas fan trips. It is not practical to design a large furnace to withstand an explosion or pressure surge caused by sudden ignition of a furnace full of flue gases.

Figure 13.2 Boiler parts hanging from the support steel.

Figure 13.3 Top support beam arrangement.

Supporting beams take pressure forces from the furnace walls, Fig. 13.4. These supporting beams encircle the furnace at regular 2–4 m intervals. The furnace wall length and width increase when the boiler is heated up. The support beam–wall

Figure 13.4 Support steel for the furnace wall.

Figure 13.5 Detail of the support structure for beams.

structure must be able to transfer thermal elongation. One way of handling this is a lug-like intermediate structure, Fig. 13.5.

Usually this wall support steel has one designated corner made weaker. This is to ensure that if there is an accident, then the force is directed to the section of the boiler house where minimum harm will occur.

13.1.2 Standing Boiler

Smaller boilers stand on boiler building floors. Package boilers and heat recovery steam generator (HRSG) boilers are

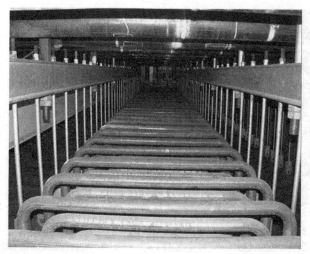

Figure 13.6 Superheater support structure.

usually of the standing type. Typically a steel frame is welded around the boiler and the whole thermal expansion takes place inside this frame.

The main problem is the transfer of horizontal thermal expansion. There must be some way other than sliding between the boiler and the floor. Rolls and plates are usually used to help with thermal expansion.

13.1.3 Superheater Support Structures

Superheater platens hang from rods, Fig. 13.6. Usually there are two rods for each platen. Rods are connected to beams. Beams are connected to larger beams. The largest beams supported by the steel structure.

13.1.4 Hanging of Drum

In large boilers the drum with the downcomers is suspended from U-type rods, Fig. 13.7. This allows for both radial and axial thermal expansion. No connecting welds are needed.

13.1.5 Hanging of Cold Surfaces

Outcoming and ingoing tubes are connected to a header. Surfaces are connected to a steel beam. The surface hangs from rods that are connected to the beams. A plate structure separates flue gas from the environment (Fig. 13.8).

Figure 13.7 Hanging of the drum.

Figure 13.8 Hanging of cold surfaces.

13.1.6 Hanging Surfaces in the Backpass

A very useful method of hanging surfaces in a backpass is to use the outlet from a low-temperature surface as a hanger tube for the rest of the surfaces. Such a tube is frequently an economizer tube.

The economizer itself hangs from steel plate supports. This kind of hanging (uncooled hanging) can be done below some 400°C. At higher temperatures the hanger tubes must be cooled (Fig. 13.9).

Figure 13.9 Installation of a furnace wall.

Figure 13.10 Platforms under construction.

13.2 Building and Platforms

To be able to access all the boiler parts and to walk around the boiler, platforms are built at suitable elevations. Platforms are mostly made from steel grille so that air, dust, dirt, etc. can move through (Fig. 13.10). Main platforms are sometimes made from concrete. Concrete stops all flows. The advantage of concrete is that it keeps the elevation below it relatively free of ash.

Platforms form the working surface in the boiler. Platforms must be built strong enough so that personnel can walk around even when maintenance to the boiler equipment is being done on the same platforms.

13.3 Insulation

All hot surfaces are insulated to ~50°C. A stainless steel plate covers the actual insulation material. The use of stainless steel plate means clean surfaces with low thermal loss. Often plastic-covered steel plates are also used for the top cover. Support beams are covered with plate, Fig. 13.11. Manholes and openings among other surfaces are not insulated.

A typical boiler wall insulation structure consists of several layers of insulation bricks. Pieces of steel wire are welded to the membrane wall. Studs (wire) support the insulation. To decrease air leakage, overlapped insulation is preferred. Insulation material is covered with a wire mesh and galvanized steel cladding. As temperatures are high, the insulation must be able to withstand 300−400°C temperatures.

Before the membrane wall construction gladding forms the pressure shield. Now the gladding thickness is typically only a few millimeters thick.

Figure 13.11 Insulation of the furnace wall, also showing a manhole.

13.4 Erection

Erection means lifting up and joining the various premanufactured pieces together. It is important for premanufactured pieces to be carefully made to predetermined measurements. These measurements should be verified before shipping. It is important to remeasure them at the building site before lifting as individual pieces tend to deform during heavy handling and they may need to be repaired before lifting and welding.

13.5 Maintenance

Often boilers start to perform less optimally when there is an issue needing maintenance. Any piece of equipment or associated structure needs to be kept in shape for proper operation to avoid losses due to poor performance and safety problems.

The boiler enclosure needs to be properly thought out so that necessary work like welding and inspection can be done without additional temporary structures. It also pays to think about how to replace heating surfaces and wall tubes (Advances in power station construction, 1986). It is necessary to think about the safety of people working inside a boiler; e.g., how to make necessary scaffolding and how to minimize exposure to falling objects. Finally, local safety authorities or the fire brigade can help in planning how to remove injured persons from premises and especially from inside the boiler through a manhole.

The boiler needs to be kept clean. During maintenance there is water washing to be done. All waste and dirty water needs to flow somewhere safely. At the start of boiler maintenance there might be ash to be displaced.

Boilers regularly need inspection. This inspection needs space. X-rays need to be examined. Teams with special equipment needs to be housed. In particular, welds and welding quality require proper space.

References

Advances in power station construction, 1986. Central Electricity Generating Board, Barnwood. Pergamon Press, Gloucester, UK, 759 p. ISBN 0080316778.

Akturk, N.U., Allan, R.E., Barrett, A.A., Brooks, W.J.D., Cooper, J.R.P., Harris, C.P., et al., 1991. In: Clapp, R.M. (Ed.), Modern Power Station Practice, vol. B, Boilers and Ancillary Plant, third ed. Pergamon press, Singapore, p. 184. ISBN 0080405126.

APPENDIX A: QUESTIONS

1. With what fuels and for which applications are BFB and CFB boilers used? List the main fuels, areas of application, and boiler size. What are the differences between BFB and CFB boilers?

2. List the main steam boiler types that are used to burn biomass in Finland.

3. List the main steam boiler types that are used to generate electricity in Europe.

4. List the type of steam boiler that would use bark (from pine and spruce) as the main fuel. Describe the boiler type, its size, and the application.

5. List the main biofuels that are used to generate electricity in Europe.

6. Why is electricity generation using steam boilers utilized around the world?

7. Describe the different kinds of biofuels available.

8. List three typical boiler types for the combustion of biofuels.

9. List and describe briefly woody biomass combustion in a large BFB.

10. List and describe briefly the main steam and water circulation types.

11. List and describe briefly the main heat transfer types used in CFB boilers.

12. List and describe briefly the main interlocks and associated measurements.

13. List and describe briefly the main boiler controls.

14. List and describe briefly the main ways to control CFB boiler operation.

15. What kinds of fuel and what kinds of big steam boiler are used in Swedish industry?

16. Describe the mechanical design of a BFB boiler furnace (walls, openings, support steel, hanging, bottom, refractory).

17. Describe how a BFB boiler furnace is built (including furnace walls, openings, support steel, hanging, bottom, refractory).

18. Describe feedwater treatment for a CFB boiler 213 kg/s, 15.7 MPa/540°C.

19. Describe the technology options to capture CO_2 from biomass burning in a CFB.

20. List the order of heat transfer surfaces and draw the surface and flue gas temperatures for a 480°C, 8.0 MPa BFB boiler that burns bark.

21. List the types of steam boiler that would use forest residue (chips, stumps, branches) as the main fuel. Describe the boiler type, its size, and the application.

22. Briefly describe the following:
    - large CFB boiler cyclone return flow superheater;
    - effect of flue gas fan efficiency on steam generation efficiency;
    - the most problematic emissions to air when burning forest residue in a large CFB boiler;
    - criteria for large CFB furnace design;
    - sootblowing in a large CFB.

23. Briefly describe the following:
    - effect of air ratio to biomass boiler flue gas CO content;
    - effect of flue gas recirculation on reducing NO_x;
    - control of steam drum water level;
    - how fuel is fed to a MSW boiler;
    - effect of reference temperature (temperature where enthalpies are 0) to boiler efficiency.

24. Briefly describe the following:
    - how to determine the furnace cross-section of a BFB;
    - effect of limestone injection on reducing SOx from CFB;
    - how fuel is fed to grate boiler burning bark;
    - effect of boiler pressure (main steam pressure) to boiler efficiency.

25. Briefly describe the following:
    - BFB flue gas air heater;
    - effect of electrostatic precipitator ash temperature to steam generation efficiency;
    - elements in bark that must be looked at for emissions
    - criteria for BFB furnace design;
    - ash removal from large CFB backpass.

26. Briefly describe the following:
    - large CFB boiler flue gas air heater;
    - effect of feedwater pump efficiency to steam generation efficiency;
    - the most problematic emissions to air when burning waste wood;
    - criteria for large CFB backpass superheater design;
    - ash removal from large CFB furnace.

27. Briefly describe the following:
    - large CFB boiler in furnace superheater;
    - effect of boiler room heating on steam generation efficiency;
    - the most problematic emissions to air when burning forest residue in a large CFB boiler;
    - criteria for large CFB boiler backpass design;
    - pressure part hanging in a large CFB.

28. Briefly describe the following:
    - large CFB boiler cyclone return flow superheater;
    - effect of auxiliary burner operation on MRC efficiency of CFB steam generation;
    - the most problematic emissions to air when burning forest residue in a large CFB boiler;
    - criteria for large CFB furnace bottom design;
    - ash removal in a large CFB.

29. Briefly describe the following:
    - Large BFB steam drum moisture separation;
    - Addition of limestone to BFB;

- the most problematic questions in fuel storage, transport and handling when burning forest residue in a large BFB boiler;
- criteria for large BFB superheater design;
- air preheating options in a large BFB.

30. List the order of heat transfer surfaces in the multifuel CFB boiler 550 MWth, 194/179 kg/s, 165/40 bar, 545/545°C and estimate the surface temperatures. Hint: first estimate the water/steam and flue gas temperatures.

31. A multifuel BFB boiler, 150 MWth, 80 bar, 482°C, needs to be retrofitted to burn wet waste. How should the heat transfer surfaces be changed?

32. Dimension roughly the furnace and the economizer of a 385 MWth, 125 MWe, 149 kg/s, 115 bar(a), 550°C MPa CFB boiler that burns bark LHV(wet) 6 MJ/kg with flue gas amount 0.60 m^3n/MJ. The furnace wall loading is 230 kW/m^2. The economizer tube outer diameter is 38.0 mm and the wall thickness is 3.2 mm. The overall heat transfer coefficient fg-eko is 67 W/m^2K and the feedwater inlet temperature is 155°C. The flue gas cp is 1.5 kJ/m^3n and the water cp is 4.0 kJ/kg.

33. Dimension roughly the furnace and the air preheater of a 385 MWth, 125 MWe, 149 kg/s, 115 bar(a), 550°C MPa CFB boiler that burns bark LHV(wet) 6 MJ/kg with flue gas amount 0.60 m^3n/MJ_{fuel}. The furnace wall loading is 230 kW/m^2. The air preheater tube outer diameter is 63.5 mm and the wall thickness is 4.0 mm. The overall heat transfer coefficient fg-ah is 27 W/m^2K and the flue gas exit temperature is 155°C. The flue gas cp is 1.5 kJ/m^3n and the volume is 0.8 m^3/kg at 0°C. The air cp is 1.3 kJ/m^3n and the volume is 0.8 m^3/kg at 0°C.

34. A multifuel BFB boiler, 150 MWth, 80 bar, 482°C, must increase the heat generated with peat from 30% to 60%. The rest of the fuel is bark. Estimate how the fuel mass flows are affected. Estimate how the flue gas flow is affected. Estimate how the boiler efficiency is affected. Estimate how the furnace outlet temperature is affected. Estimate how the superheating is affected. Estimate how the boiler flue gas exit temperature is affected.

		Peat	Bark
Moisture	%	40.0	55.0
Ash	%	8.9	3.9
LHV(af)	MJ/kg	10.8	7.1
C	% (daf)	55.8	53.6
H	% (daf)	5.9	6.2
O	% (daf)	36.0	39.7
S	% (daf)	0.24	0.04
N	% (dry)	2.0	0.46
Cl	% (daf)	0.05	0.02

35. Dimension roughly the furnace and the economizer of a 150 MWth, 80 bar, 482°C, multifuel BFB boiler burning bark LHV(wet) 6 MJ/kg with flue gas amount 0.60 m^3n/MJ. The furnace wall loading is 230 kW/m^2. The economizer tube outer diameter is 38.0 mm and the wall thickness is 3.2 mm. The overall heat transfer coefficient fg-eko is 67 W/m^2K and the feedwater inlet temperature is 155°C. The flue gas cp is 1.5 kJ/m^3n and the water cp is 4.0 kJ/kg.

		Peat	Bark
Moisture	%	40.0	55.0
Ash	%	8.9	3.9
LHV(af)	MJ/kg	10.8	7.1
C	% (daf)	55.8	53.6
H	% (daf)	5.9	6.2
O	% (daf)	36.0	39.7
S	% (daf)	0.24	0.04
N	% (dry)	2.0	0.46
Cl	% (daf)	0.05	0.02

36. Dimension roughly the furnace and the air preheater of a 150 MWth, 80 bar, 482°C, multifuel BFB boiler burning bark LHV(wet) 6 MJ/kg with flue gas amount 0.60 m^3n/MJ. The furnace wall loading is 230 kW/m^2. The air preheater tube outer diameter is 55.0 mm and the wall thickness is 2.0 mm. The overall heat transfer coefficient fg-aph is 27 W/m^2K and the air inlet temperature is 35°C. The flue gas cp is 1.5 kJ/m^3n and the air cp is 1.0 kJ/kg.

		Peat	Bark
Moisture	%	40.0	55.0
Ash	%	8.9	3.9
LHV(af)	MJ/kg	10.8	7.1
C	% (daf)	55.8	53.6
H	% (daf)	5.9	6.2
O	% (daf)	36.0	39.7
S	% (daf)	0.24	0.04
N	% (dry)	2.0	0.46
Cl	% (daf)	0.05	0.02

37. What are the fundamental subjects to be considered in the design of the backpass evaporating surface in natural circulation systems?

38. Describe a large CFB boiler fuel system (equipment, purpose, control, measurements) when the fuels burned are coal, peat, stumps, and bark.

39. Describe the fuel and ash handling system of a large utility CFB boiler that burns coal, bark, peat, and wood waste.

40. Describe the differences in fuel handling systems for different kinds of biofuel used in Scandinavia.

41. Describe a large CFB boiler air system (equipment, purpose, control, measurements).

42. Describe a large BFB boiler air system (equipment, purpose, control, measurements).

43. What are the fundamental subjects to be considered in the design of the steam–water flow system in a small BFB boiler?

44. Describe the biomass fuel handling for a CFB 213 kg/s, 15.7 MPa/540°C, boiler that uses bark, chips, peat, stumps, and branches.

45. What are the critical emissions and their relevant limits when firing peat, bark, and coal in a CFB 213 kg/s, 15.7 MPa/540°C boiler.

46. What are the emissions and why do they cause concern when firing peat, bark, and coal in a CFB 213 kg/s, 15.7 MPa/540°C boiler?

47. Describe the combined heat and power generation from biomass in the EU, including technology used, challenges, advantages, and threats.

48. Describe the erection sequence of a large biomass boiler.

49. Describe the fly ash formation from biomass burning in a CFB.

50. Compare CFB boiler design with natural circulation and once-through design. How does this affect furnace design? How does this affect superheater design? When would once-through be better? When would natural circulation be better?

51. Describe the typical corrosion and erosion considerations in a big biomass CFB boiler backpass.

52. What kinds of emission does one get when burning biofuels (wood, agri, and waste) in 10–130 MWth grate-BFB and -CFB boilers in Sweden?

53. What are the fundamental subjects that should be considered in the thermal and hydraulic design of water tube systems in once-through boilers?

54. What is the normal operating window of the once-through boiler turbine load?

55. Define the typical inclination angles of the tubes used in conventional once-through boilers.

56. Define the level of the fluid flow pressure drop of the once-through boiler.

57. Define the types of evaporator tube that could be used for a forced circulation boiler case.

58. What are the fundamental subjects to be considered in the design of flow systems in forced circulation systems?

59. Compare the BFB boiler layout design with biomass burning. How does size affect surface placement? How does fuel affect layout design? What things determine which design is better?

60. Compare large BFB boiler backpass heat transfer surface design with biomass burning and waste burning. How does this affect air heater design? How does this affect superheater design? What things determine which design is better?

61. Compare BFB boiler heat transfer surface construction, hanging, and insulation design with biomass burning. How does boiler size affect mechanical design? How does this affect furnace design? What things determine which design is better?

62. Compare large CFB boiler design with coal and biomass (wood chips). How does this affect furnace design? How does this affect superheater design? What are the effects on size, cost, and efficiency?

APPENDIX B: H-S DIAGRAM

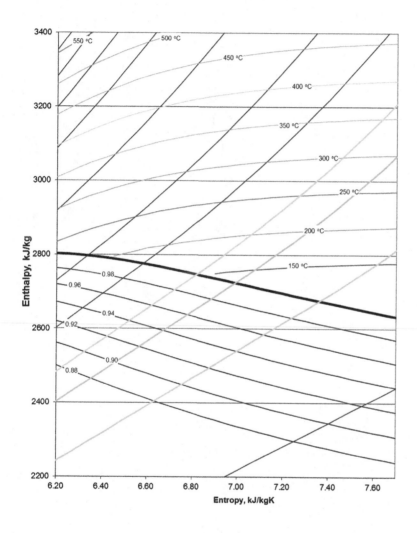

APPENDIX C: STEAM TABLES

State of saturation (Temperature table)

Saturated Temperature °C	Saturated Pressure MPa	Specific Volume Liquid m³/kg	Specific Volume Vapor m³/kg	Specific Enthalpy Liquid kJ/kg	Specific Enthalpy Vapor kJ/kg	Specific Entropy Liquid kJ/kgK	Specific Entropy Vapor kJ/kgK
5.00	0.00087	0.001000	147.0	21.0	2510.1	0.1059	8.975
10.00	0.00123	0.001000	106.31	42.0	2519.2	0.1060	8.975
15.00	0.00171	0.001001	77.88	63.0	2528.4	0.1061	8.975
20.00	0.00234	0.001002	57.76	83.9	2537.5	0.1063	8.974
25.00	0.00317	0.001003	43.34	104.8	2546.5	0.1066	8.974
30.00	0.00425	0.001004	32.88	125.7	2555.6	0.1069	8.973
35.00	0.00563	0.001006	25.21	146.6	2564.6	0.1072	8.973
40.00	0.00738	0.001008	19.517	167.5	2573.5	0.1076	8.972
50.00	0.01235	0.001012	12.028	209.3	2591.3	0.1085	8.971
60.00	0.01995	0.001017	7.668	251.2	2608.8	0.1096	8.969
70.00	0.03120	0.001023	5.040	293.0	2626.1	0.1108	8.967
80.00	0.04741	0.001029	3.405	334.9	2643.0	0.1122	8.964
90.00	0.07018	0.001036	2.359	377.0	2659.5	0.1136	8.962
100.00	0.10142	0.001043	1.6719	419.1	2675.6	0.1152	8.959
110.00	0.14338	0.001052	1.2094	461.4	2691.1	0.1169	8.957
120.00	0.19867	0.001060	0.8913	503.8	2705.9	0.1187	8.954
130.00	0.27026	0.001070	0.6681	546.4	2720.1	0.1207	8.950
140.00	0.36150	0.001080	0.5085	589.2	2733.4	0.1227	8.947
150.00	0.47610	0.001091	0.3925	632.3	2745.9	0.1249	8.943
160.00	0.61814	0.001102	0.3068	675.6	2757.4	0.1272	8.939
170.00	0.79205	0.001114	0.24262	719.2	2767.9	0.1296	8.935
180.00	1.00263	0.001127	0.19386	763.2	2777.2	0.1322	8.931
200.00	1.55467	0.001157	0.12722	852.4	2792.1	0.1378	8.922
220.00	2.31929	0.001190	0.08610	943.6	2801.1	0.1441	8.911
240.00	3.34665	0.001229	0.05971	1037.5	2803.1	0.1513	8.899
260.00	4.69207	0.001276	0.04218	1134.8	2796.6	0.1596	8.886
280.00	6.41646	0.001333	0.03015	1236.7	2779.8	0.1692	8.870
300.00	8.58771	0.001404	0.02166	1344.8	2749.6	0.1808	8.851
320.00	11.28386	0.001499	0.01548	1462.1	2700.7	0.1954	8.827
340.00	14.60018	0.001638	0.010784	1594.4	2622.1	0.2153	8.795
360.00	18.66640	0.001895	0.006945	1761.5	2481.0	0.2482	8.742
370.00	21.04337	0.002222	0.004946	1892.6	2333.5	0.2847	8.685

State of saturation (Pressure table)

Saturated Pressure MPa	Saturated Temperature °C	Specific Volume Liquid m³/kg	Specific Volume Vapor m³/kg	Specific Enthalpy Liquid kJ/kg	Specific Enthalpy Vapor kJ/kg	Specific Entropy Liquid kJ/kgK	Specific Entropy Vapor kJ/kgK
0.00100	6.97	0.001000	129.2	29.3	2513.7	0.1059	8.975
0.00200	17.50	0.001001	66.99	73.4	2532.9	0.1062	8.974
0.00300	24.08	0.001003	45.66	101.0	2544.9	0.1065	8.974
0.00400	28.96	0.001004	34.79	121.4	2553.7	0.1068	8.973
0.00500	32.88	0.001005	28.19	137.8	2560.8	0.1071	8.973
0.00700	39.00	0.001007	20.53	163.4	2571.8	0.1075	8.972
0.01000	45.81	0.001010	14.67	191.8	2583.9	0.1081	8.971
0.02000	60.06	0.001017	7.648	251.4	2608.9	0.1096	8.969
0.03000	69.10	0.001022	5.229	289.2	2624.6	0.1107	8.967
0.04000	75.86	0.001026	3.993	317.6	2636.1	0.1116	8.965
0.05000	81.32	0.001030	3.240	340.5	2645.2	0.1124	8.964
0.07000	89.93	0.001036	2.365	376.7	2659.4	0.1136	8.962
0.10000	99.61	0.001043	1.694	417.4	2674.9	0.1151	8.959
0.20000	120.21	0.001061	0.8857	504.7	2706.2	0.1188	8.953
0.30000	133.53	0.001073	0.6058	561.5	2724.9	0.1214	8.949
0.40000	143.61	0.001084	0.4624	604.7	2738.1	0.1235	8.946
0.50000	151.84	0.001093	0.3748	640.2	2748.1	0.1253	8.943
0.70000	164.95	0.001108	0.2728	697.1	2762.7	0.1284	8.937
1.00000	179.89	0.001127	0.1943	762.7	2777.1	0.1322	8.931
1.50000	198.30	0.001154	0.1317	844.7	2791.0	0.1373	8.923
2.00000	212.38	0.001177	0.09958	908.6	2798.4	0.1416	8.915
3.00000	233.86	0.001217	0.06666	1008.4	2803.3	0.1490	8.903
4.00000	250.36	0.001253	0.04978	1087.4	2800.9	0.1554	8.893
6.00000	275.59	0.001319	0.03245	1213.7	2784.6	0.1669	8.874
8.00000	295.01	0.001385	0.02353	1317.1	2758.6	0.1777	8.856
10.00000	311.00	0.001453	0.01803	1407.9	2725.5	0.1884	8.839
12.00000	324.68	0.001526	0.01427	1491.3	2685.6	0.1995	8.821
14.00000	336.67	0.001610	0.01149	1570.9	2638.1	0.2114	8.801
16.00000	347.36	0.001710	0.00931	1649.7	2580.8	0.2250	8.780
18.00000	356.99	0.001840	0.007499	1732.0	2509.5	0.2416	8.753
20.00000	365.75	0.002039	0.005858	1827.1	2411.4	0.2649	8.716
22.00000	373.71	0.002750	0.003577	2021.9	2164.2	0.3340	8.608

Enthalpy (kJ/kg) as function of temperature (°C) and pressure (MPa)

Temperature °C	Pressure, MPa							
	4.0	6.0	8.0	10.0	12.0	14.0	16.0	20.0
20.0	87.7	89.5	91.4	93.3	95.2	97.0	98.9	102.6
60.0	254.5	256.2	257.8	259.5	261.2	262.9	264.5	267.9
100.0	422.0	423.5	425.0	426.5	428.1	429.6	431.1	434.1
140.0	591.6	592.9	594.2	595.5	596.8	598.1	599.5	602.1
160.0	677.6	678.7	679.9	681.1	682.3	683.5	684.7	687.2
180.0	764.7	765.7	766.8	767.8	768.9	769.9	771.0	773.2
200.0	853.4	854.2	855.1	855.9	856.8	857.7	858.6	860.4
220.0	944.1	944.7	945.3	945.9	946.5	947.2	947.8	949.2
240.0	1037.6	1037.8	1038.0	1038.3	1038.6	1039.0	1039.3	1040.1
260.0		1134.6	1134.3	1134.1	1134.0	1133.9	1133.8	1133.8
280.0			1235.8	1234.8	1233.9	1233.1	1232.5	1231.3
300.0				1343.1	1340.9	1339.0	1337.2	1334.1
320.0					1460.3	1455.9	1451.9	1445.3
340.0							1587.3	1571.5
360.0								1740.0
260.0	2837.2							
280.0	2902.9	2805.2						
300.0	2961.7	2885.5	2786.4					
340.0	3068.1	3014.9	2953.9	2882.1	2793.5	2672.4		
360.0	3118.1	3072.0	3020.6	2962.6	2895.9	2816.4	2715.6	
360.0	3118.1	3072.0	3020.6	2962.6	2895.9	2816.4	2715.6	
380.0	3166.7	3126.1	3081.8	3033.1	2979.1	2918.3	2848.3	2659.2
400.0	3214.4	3178.2	3139.3	3097.4	3051.9	3002.2	2947.5	2816.8
420.0	3261.4	3228.8	3194.2	3157.5	3118.2	3076.1	3030.9	2928.5
440.0	3307.9	3278.3	3247.3	3214.6	3180.1	3143.6	3105.0	3020.3
460.0	3354.0	3327.0	3298.9	3269.5	3238.8	3206.7	3173.0	3100.6
480.0	3400.0	3375.2	3349.5	3322.9	3295.2	3266.5	3236.7	3173.4
500.0	3445.8	3422.9	3399.4	3375.1	3350.0	3324.1	3297.3	3241.2
520.0	3491.6	3470.4	3448.6	3426.3	3403.4	3379.8	3355.6	3305.2
540.0	3537.3	3517.6	3497.5	3476.9	3455.8	3434.2	3412.1	3366.4
560.0	3583.1	3564.7	3546.0	3526.9	3507.4	3487.5	3467.3	3425.6
580.0	3628.9	3611.8	3594.3	3576.5	3558.4	3540.1	3521.4	3483.0
600.0	3674.8	3658.8	3642.4	3625.8	3609.0	3591.9	3574.6	3539.2

Specific volume (m³/kg) as function of temperature (°C) and pressure (MPa)

Temperature °C	Pressure, MPa						
	6.0	8.0	10.0	12.0	14.0	16.0	20.0
20.0	0.0009991	0.0009982	0.0009973	0.0009964	0.0009955	0.0009947	0.0009929
60.0	0.0010144	0.0010136	0.0010127	0.0010118	0.0010109	0.0010101	0.0010084
100.0	0.0010405	0.0010395	0.0010385	0.0010375	0.0010365	0.0010356	0.0010337
140.0	0.0010762	0.0010750	0.0010738	0.0010726	0.0010714	0.0010702	0.0010679
160.0	0.0010981	0.0010967	0.0010954	0.0010940	0.0010926	0.0010913	0.0010886
180.0	0.0011232	0.0011216	0.0011200	0.0011184	0.0011168	0.0011152	0.0011122
200.0	0.0011521	0.0011501	0.0011482	0.0011463	0.0011444	0.0011426	0.0011390
220.0	0.0011856	0.0011832	0.0011808	0.0011785	0.0011763	0.0011740	0.0011697
240.0	0.0012253	0.0012222	0.0012192	0.0012163	0.0012135	0.0012107	0.0012053
260.0	0.0012734	0.0012693	0.0012653	0.0012615	0.0012578	0.0012542	0.0012472
280.0		0.0013282	0.0013226	0.0013173	0.0013121	0.0013072	0.0012978
300.0			0.0013980	0.0013898	0.0013820	0.0013746	0.0013611
320.0				0.0014937	0.0014797	0.0014671	0.0014449
340.0						0.0016163	0.0015693
360.0							0.0018245
260.0							
280.0	0.03320						
300.0	0.03619	0.02428					
340.0	0.04114	0.02899	0.02149	0.01621	0.01200		
360.0	0.04334	0.03092	0.02333	0.01812	0.01423	0.01106	
360.0	0.04334	0.03092	0.02333	0.01812	0.01423	0.01106	
380.0	0.04542	0.03268	0.02495	0.01971	0.01587	0.01288	0.00826
400.0	0.04742	0.03435	0.02644	0.02111	0.01724	0.01428	0.00995
420.0	0.04936	0.03593	0.02783	0.02239	0.01846	0.01548	0.01120
440.0	0.05124	0.03745	0.02915	0.02358	0.01958	0.01655	0.01225
460.0	0.05308	0.03893	0.03041	0.02471	0.02062	0.01753	0.01317
480.0	0.05489	0.04036	0.03163	0.02579	0.02160	0.01845	0.01401
500.0	0.05667	0.04177	0.03281	0.02683	0.02255	0.01932	0.01479
520.0	0.05843	0.04315	0.03397	0.02784	0.02345	0.02016	0.01553
540.0	0.06016	0.04450	0.03510	0.02882	0.02433	0.02096	0.01623
560.0	0.06188	0.04584	0.03621	0.02978	0.02519	0.02174	0.01690
580.0	0.06358	0.04716	0.03730	0.03072	0.02602	0.02250	0.01755
600.0	0.06526	0.04846	0.03838	0.03165	0.02684	0.02324	0.01818

Thermal conductivity, W/mK as function of temperature °C and pressure MPa

Temperature °C	Pressure, MPa							
	4.0	6.0	8.0	10.0	12.0	14.0	16.0	20.0
20.0	0.6016	0.6027	0.6038	0.6048	0.6059	0.6070	0.6080	0.6101
60.0	0.6528	0.6539	0.6549	0.6559	0.6569	0.6579	0.6589	0.6610
100.0	0.6799	0.6810	0.6821	0.6832	0.6843	0.6854	0.6865	0.6886
140.0	0.6872	0.6885	0.6897	0.6910	0.6922	0.6934	0.6947	0.6971
160.0	0.6842	0.6856	0.6869	0.6883	0.6896	0.6910	0.6923	0.6950
180.0	0.6769	0.6785	0.6800	0.6815	0.6830	0.6845	0.6860	0.6889
200.0	0.6655	0.6673	0.6690	0.6707	0.6724	0.6740	0.6757	0.6789
220.0	0.6499	0.6519	0.6539	0.6558	0.6577	0.6596	0.6615	0.6652
240.0	0.6297	0.6321	0.6344	0.6367	0.6389	0.6411	0.6433	0.6475
260.0		0.6074	0.6102	0.6130	0.6157	0.6183	0.6209	0.6258
280.0			0.5805	0.5840	0.5873	0.5905	0.5936	0.5997
300.0				0.5481	0.5525	0.5567	0.5607	0.5683
320.0					0.5087	0.5146	0.5202	0.5304
340.0							0.4677	0.4831
360.0								0.4198
260.0	0.05030							
280.0	0.04967	0.05902						
300.0	0.05014	0.05665	0.06724					
340.0	0.05272	0.05666	0.06209	0.06976	0.08136	0.10163		
360.0	0.05443	0.05769	0.06198	0.06772	0.07568	0.08736	0.10638	
360.0	0.05443	0.05769	0.06198	0.06772	0.07568	0.08736	0.10638	
380.0	0.05630	0.05903	0.06241	0.06667	0.07213	0.07937	0.08941	0.12939
400.0	0.05836	0.06080	0.06372	0.06725	0.07153	0.07682	0.08345	0.10335
420.0	0.06050	0.06271	0.06529	0.06830	0.07185	0.07605	0.08107	0.09464
440.0	0.06273	0.06476	0.06708	0.06974	0.07279	0.07631	0.08040	0.09077
460.0	0.06501	0.06691	0.06904	0.07143	0.07414	0.07720	0.08067	0.08914
480.0	0.06736	0.06914	0.07112	0.07332	0.07576	0.07849	0.08153	0.08873
500.0	0.06975	0.07144	0.07330	0.07534	0.07758	0.08005	0.08278	0.08910
520.0	0.07219	0.07380	0.07556	0.07747	0.07955	0.08183	0.08431	0.08997
540.0	0.07467	0.07621	0.07788	0.07969	0.08164	0.08376	0.08604	0.09119
560.0	0.07719	0.07868	0.08027	0.08198	0.08383	0.08581	0.08794	0.09268
580.0	0.07975	0.08118	0.08271	0.08434	0.08609	0.08796	0.08996	0.09437
600.0	0.08234	0.08372	0.08519	0.08676	0.08842	0.09020	0.09209	0.09622

Dynamic viscosity (10^{-3} Pa s) as function of temperature (°C) and pressure (MPa)

Temperature °C	Pressure, MPa							
	4.0	6.0	8.0	10.0	12.0	14.0	16.0	20.0
20.0	1.0000	0.9992	0.9984	0.9977	0.9970	0.9963	0.9956	0.9944
60.0	0.4673	0.4677	0.4682	0.4686	0.4691	0.4696	0.4701	0.4710
100.0	0.2828	0.2833	0.2839	0.2844	0.2849	0.2855	0.2860	0.2871
140.0	0.1975	0.1980	0.1985	0.1990	0.1995	0.2000	0.2005	0.2015
160.0	0.1711	0.1716	0.1721	0.1726	0.1731	0.1736	0.1740	0.1750
180.0	0.1509	0.1514	0.1519	0.1524	0.1528	0.1533	0.1538	0.1548
200.0	0.1349	0.1354	0.1359	0.1364	0.1369	0.1374	0.1379	0.1388
220.0	0.1220	0.1225	0.1230	0.1235	0.1240	0.1245	0.1249	0.1259
240.0	0.1110	0.1116	0.1121	0.1126	0.1131	0.1137	0.1142	0.1152
260.0		0.1021	0.1026	0.1032	0.1038	0.1043	0.1049	0.1059
280.0			0.0940	0.0947	0.0953	0.0959	0.0965	0.0977
300.0				0.0865	0.0872	0.0880	0.0887	0.0901
320.0					0.0788	0.0798	0.0807	0.0825
340.0							0.0716	0.0742
360.0								0.0628
260.0	0.0180							
280.0	0.01894	0.01873						
300.0	0.01988	0.01973	0.01965					
340.0	0.02171	0.02164	0.02162	0.02167	0.02185	0.02229		
360.0	0.02261	0.02256	0.02257	0.02263	0.02277	0.02305	0.02357	
360.0	0.02261	0.02256	0.02257	0.02263	0.02277	0.02305	0.02357	
380.0	0.02349	0.02348	0.02350	0.02356	0.02370	0.02391	0.02426	0.02578
400.0	0.02437	0.02437	0.02441	0.02449	0.02461	0.02480	0.02508	0.02603
420.0	0.02524	0.02526	0.02531	0.02540	0.02552	0.02570	0.02594	0.02667
440.0	0.02610	0.02614	0.02620	0.02629	0.02642	0.02659	0.02680	0.02742
460.0	0.02696	0.02701	0.02708	0.02718	0.02730	0.02747	0.02767	0.02821
480.0	0.02781	0.02787	0.02794	0.02805	0.02818	0.02834	0.02853	0.02903
500.0	0.02865	0.02872	0.02880	0.02891	0.02904	0.02920	0.02939	0.02985
520.0	0.02949	0.02956	0.02965	0.02976	0.02990	0.03005	0.03023	0.03067
540.0	0.03031	0.03039	0.03049	0.03061	0.03074	0.03090	0.03107	0.03149
560.0	0.03114	0.03122	0.03133	0.03144	0.03158	0.03173	0.03190	0.03231
580.0	0.03195	0.03204	0.03215	0.03227	0.03241	0.03256	0.03273	0.03312
600.0	0.03276	0.03286	0.03297	0.03309	0.03323	0.03338	0.03354	0.03392

INDEX

Note: Page numbers followed by "*f*" refer to figures.

Printed in the United States
By Bookmasters